CATHERINE CAUFIELD

MULTIPLE EXPOSURES

Chronicles of the Radiation Age

The University of Chicago Press

To Philip Williams

The University of Chicago Press, Chicago 60637
© 1989 by Catherine Caufield
All rights reserved. Originally published 1989
University of Chicago Press edition 1990
Printed in the United States of America

99 98 97 96 95 94 93 92 91 90 54321

Published by arrangement with Harper & Row, Publishers, Inc.

Library of Congress Cataloging-in-Publication Data

Caufield, Catherine.
    Multiple exposures : chronicles of the radiation age / Catherine
Caufield.—University of Chicago Press ed.
        p.   cm.
    First published by Secker, c1989.
    Includes bibliographical references.
    ISBN 0-226-09785-4 (alk. paper)
    1. Ionizing radiation—Dosage—Standards—History.   2. Ionizing
radiation—Safety measures—History.   3. Radiation injuries—
History.   I. Title.
RA569.C38   1990
363.17'99—dc20   89-20637
                                                               CIP

This book is printed on acid-free paper.

Contents

Illustrations

The author and publishers are grateful to the following for permission to reproduce photographs: Dr Richard Mould, Nos 1, 2, 3, 4, 7, 8, 9; Topham Picture Library, Nos 11, 12, 13, 14, 15; BBC Hulton Picture Library, Nos 5, 6, 10 (The Bettman Archive).

Acknowledgements

I am more than grateful to the many kind people, from busy academics to wary industrialists, from understaffed environmental organizations to bereaved families, who helped me during the research and writing of this book. I thank everyone who granted me an interview, answered a letter, provided me with documents, or spent time on the telephone with me. Besides those people mentioned in the text, I would like to thank Robert Alvarez, Saul Bloom, Stuart Boyle, Gray Brechin, Jeremy Cherfas, John Cobb, Dianne D'Arrigo, David Dawdy, Roland Firston, John Gofman, Janet Gordon, Robin Grove-White, Patrick Green, Barton Hacker, Jim Harding, David Kaplan, Nan Hearst, Sally Hughes, William Lawless, Katherine Rose Lewis, Walter Patterson, David Pesonen, Dale Preston, William Schull, William Shawn, Chris Shuey, Lauriston Taylor, Sidney Wolfe, and Brian Wynne.

Preface

There was a time when a person with an interest in science might have been able to keep up with the frontiers of discovery. Today, such a feat is virtually impossible because science itself has become so divided that, even within a single field, many experts are isolated from one another by their own specializations.

Unable to understand the technology that affects much of our lives, we are increasingly reliant on the various specialists to control and interpret these technologies for us. There are many different reactions to this enforced dependence: some people are eager to believe that they are in safe hands; others are resentful at losing control and distrustful of those who have assumed it.

Unfortunately, scientists have all too often tried to gain the public's confidence by making misleading statements. Part of the trouble is that scientists think differently from the rest of us. To a scientist there is no absolute certainty, no immutable fact – only probabilities and an evolving understanding of how the world works. A simple statement, such as 'this X-ray is perfectly safe' or 'radiation levels here are within the internationally-accepted safety limits', may mean very different things to the layperson and the scientist. A scientist will recognize the unspoken provisos in the statements and know that a more accurate rendering would be 'as far as we know, this X-ray has a very low probability of causing you detectable harm', and 'due to technical and financial restrictions, we have not actually measured radiation levels in the area, but our computer programs indicate that radiation levels will be within the international safety limits'. To a layman, of course, the original statements sound very much like guarantees or promises. When, as they often have been, these 'promises' are broken, the public begins to develop a distrust of science and scientists – a distrust that may grow to irrational proportions.

Mankind's brief acquaintance with ionizing radiation* seems almost to have been designed to exacerbate these feelings of powerlessness and suspicion. X-rays, radium, nuclear fission – each new discovery was greeted with wild enthusiasm, which gave way to alarm when unforeseen side-effects appeared. Protection measures were introduced, and always, sooner or later, they had to be strengthened, and strengthened again. This repetitive cycle has made many people suspicious of current radiation protection standards: why trust something that has always proved to need revision? The public has also been made cynical by the fact that, for reasons ranging from ignorance to national security, information about nuclear matters is often entangled in a web of secrecy, misinformation, and lies. Those who study the subject more closely are also surprised by the extent to which radiation protection standards are not based on scientific certainty, but on judgement, hunches, and compromise.

And yet, the health and safety regulations governing the use of ionizing radiation, not just in the wealthy industrial countries, but worldwide, are far beyond those achieved by any other industry. Ionizing radiation is the best-understood and most tightly-controlled toxic substance known to man. This does not mean, however, that ionizing radiation is safe or that the current protection standards are adequate. There are many still-unanswered questions about the genetic effects of radiation, about its behaviour at very low doses, and about its link to diseases other than cancer. Today, the threat of ionizing radiation comes not so much from dramatically high individual doses, but from the growing dose to the entire population due to radiation's ever-increasing popularity as a medical and industrial tool.

The story of our efforts to use and protect ourselves from ionizing radiation raises a number of interesting questions. Are we delegating too much power and responsibility to technical experts? Are scientists honest about the limitations of their calling? Are our demands – for safety, for certainty, for all the benefits and none of the risks of technology – realistic? And, above all for many people, how well are we protecting ourselves and our descendants?

*This book is about radiations at the high-energy end of the electromagnetic radiation spectrum. These high-energy, or ionizing, radiations include X-rays, gamma rays, neutrons, and alpha and beta rays. The low-energy, or non-ionizing end of the spectrum includes ultraviolet, visible, and infra-red light, microwaves, radio waves, and the extremely low frequency radiations emitted by electric powerlines and video display terminals.

PART ONE

CHAPTER ONE

Discovery

At first he told no one. For almost two months, Wilhelm Roentgen ate and slept in his laboratory at the University of Würzburg, working doggedly to make sense of the strange thing he had seen. His colleagues in the department of physics were curious, but Roentgen was tantalizingly silent. To a close friend who queried his unusual behaviour he said only, 'I have discovered something interesting.'

It began on 8 November 1895, as he was experimenting with a glass bulb from which most of the air had been removed, an apparatus known as a Crooke's tube. In his darkened laboratory, Roentgen passed an electric current through the tube, which was wrapped in lightproof paper. Suddenly he was astonished by a soft glow at the other end of the room. He switched off the current, and the glow disappeared. Excited and confused, he turned the current on again. Again something shone in the darkness. He lit a match and saw that the glow came from a screen he had left out on a table. The screen was coated with barium platino-cyanide, a chemical compound that gives off light when its atoms are energized. Over and over again Roentgen repeated the experiment, always with the same result: the screen glowed in response to some invisible energy radiating from the tube. But what kind of energy?

So absorbed was Roentgen in his work that Friday night that he lost track of time and of his surroundings. The caretaker knocked at the door, came in, looked around for a piece of equipment, and left without Roentgen's having noticed him. Frau Roentgen dispatched a servant to call her husband to dinner several times before he finally appeared. He sat at the table in silence, not seeming to hear his wife's questions, ate little, and rushed back to his laboratory.

In the following weeks, Roentgen worked night and day on his discovery. He found that the invisible radiation passed through paper, copper, and other substances that block visible light, but that human

3

bones, certain metals, and even ordinary glass were relatively opaque to it. He named the new rays X, for mystery.

Placing various objects between the source of the rays and a photographic plate, Roentgen was able to take X-ray pictures. For one of the first, he asked his wife to place her hand on a photographic plate while he trained the rays on to it for 15 minutes. The picture, which still exists, shows only the bones of her hand and her wedding ring; the skin and soft tissues did not show up on the plate because the rays passed right through them. According to Roentgen's biographer, Otto Glasser, 'She could hardly believe that this bony hand was her own and shuddered at the thought that she was seeing her skeleton. To Mrs Roentgen, as to many others later, this experience gave a vague premonition of death.'

Roentgen's feverish study of X-rays was rewarded by the incredible speed with which his work was publicly acknowledged. In the last days of December he submitted a paper to the secretary of the Würzburg Physical-Medical Society, and persuaded him to take the unusual step of publishing it before it had been read at a meeting of the society. The paper, 'On a New Kind of Ray', appeared in the society's *Proceedings*, dated 28 December 1895. By New Year's Day 1896, Roentgen had reprints, which he mailed with sample X-ray photographs to a number of colleagues.

News of Roentgen's sensational discovery swept through the popular press before his scientific paper was available. The Vienna *Presse* broke the story on Sunday 5 January with a front-page article. The Vienna correspondent of London's *Daily Chronicle* wired home a piece that appeared the following day in both the *Chronicle* and the *New York Sun*. It read, in part:

> The noise of war's alarms should not distract attention from the marvellous triumph of science which is reported from Vienna. It is announced that Professor Routgen [sic] of the Wurzburg University has discovered a light which, for the purposes of photography, will penetrate wood, flesh, and most other organic substances. The professor has succeeded in photographing metal weights which were in a closed wooden case; also a man's hand, which shows only the bones, the flesh being invisible.

On 16 January the *New York Times*, still misspelling Roentgen's name, reported that, 'Men of science in this city are awaiting with the utmost impatience the arrival of European technical journals which will give them the full particulars of Professor Routgen's great discovery.'

When he delivered his first public lecture on X-rays, at Würzburg on 23 January 1896, Roentgen spoke to a packed auditorium. Afterwards he asked Albert von Kolliker, an anatomist at the university, to submit to an X-ray photograph on stage. Kolliker put his hand on a photographic plate and Roentgen passed a current through a blacked-out glass tube. When the exposed film clearly showing the bones of Kolliker's hand was held aloft, the audience erupted in cheers, and Kolliker made an emotional proposal that the new phenomenon be called Roentgen rays. The name was widely used for a time, but never by Roentgen himself.

One of the first people to take up X-rays was Thomas Alva Edison, already famous for his work on electric light and power, moving pictures, and the phonograph. His passion for technology was combined with a showmanship that kept the press and the public fascinated by anything he did. On 5 February 1896 William Randolph Hearst, another arch-publicist, knowing of Edison's interest, telegraphed him to ask a seemingly impossible task of the infant technology:

WILL YOU AS AN ESPECIAL FAVOUR TO THE JOURNAL UNDERTAKE
TO MAKE CATHODOGRAPH OF HUMAN BRAIN KINDLY TELEGRAPH
ANSWER AT OUR EXPENSE.

Edison agreed, and the next day reporters from several papers staked out his laboratory in West Orange, New Jersey, waiting for the first X-ray of the living human brain. 'For three weeks, more than twenty newspaper reporters were stationed at the laboratory, the work going on nights, days, and Sundays,' Edison's assistant William Meadowcraft later recalled. The *Electrical Review*, reporting on the experiment, characterized Edison as 'a man who knows nothing of the passing of day and night'. Edison missed his first self-imposed deadline for the X-ray on 8 February, but Micawber-like, continued to assure reporters that success was imminent. Day by day, newspapers as far away as Nevada and Washington State regaled their readers with the story of Edison's quest, of his many reverses, and of his expectations of triumph in the near future. But after weeks of disappointment the band of reporters dwindled and finally faded away. Newspaper editors turned to other things, and so did Edison.

His public failure did not sour Edison on X-rays. He and his many assistants continued to work on improving X-ray images. He developed one of the first fluoroscopes, a device for obtaining instantaneous X-ray images by projecting the rays through an object on to a fluorescent screen instead of on to a photographic plate. By coating the screen with calcium tungstate, instead of the barium platino-cyanide that Roentgen had used, Edison achieved images six times brighter than Roentgen's. He also enclosed the screen in a convenient hand-held device reminiscent of the then-popular stereoscope, a move that made X-rays, in today's jargon, 'user-friendly' – and dangerous, since it encouraged people to regard fluoroscopes as toys. Edison was not only an inventor, but also a businessman who knew how to create a demand for his inventions. In May 1896 he exhibited his improved fluoroscope at the Electric Light Association Exposition at Grand Central Palace in New York. It was the first public display of X-rays in America and possibly in the world. At this demonstration, and the hundreds that followed around the country, thousands of people queued to put their hands, legs, or heads into the path of the rays and watch themselves on a fluorescent screen. By 25 March Edison had licensed the mass-production and sale of his fluoroscope, and shortly thereafter a complete 'Thomas A. Edison X-ray kit' was on the market.

Within weeks of the discovery of X-rays, Europe and North America were in the grip of what the *Electrical Review* called 'the Roentgen mania'. A farmer in Iowa claimed that he had used X-rays to transmute a worthless piece of metal to pure gold with a value of $153. Frances Willard, the president of the Women's Christian Temperance Union, was hopeful that drunkards and smokers would swear off their bad habits when X-rays enabled them to see how they were damaging themselves. X-rays of the head, it was suggested, might cure criminal behaviour. One newspaper asserted that, 'At the College for Physicians and Surgeons, the Roentgen rays were used to reflect anatomic diagrams directly into the brains of the students, making a much more enduring impression than the ordinary methods of learning anatomical details.' An experimenter claimed in a scientific journal to have made a dog salivate by projecting an X-ray image of a bone on to his brain.

Not everyone was thrilled by the power of X-rays. Thurston Holland, one of the first radiologists in England, feared he had 'done someone a very bad turn' when, after a lecture which included the information that diamonds are transparent to X-rays, he invited members of the audience to see their hands on a fluorescent screen. Upon finding that the diamonds in her large ring were opaque, 'a very overdressed lady made quite unprintable and vitriolic remarks,' recalled a rueful Holland.

Ribald jokes abounded about the possibilities of using X-rays to undress ladies on the street or on stage. A poem in the American magazine *Photography* concluded thus,

> I'm full of daze
> Shock and amaze;
> For now-a-days
> I hear they'll gaze
> Thru' cloak and gown – and even stays,
> These naughty, naughty Roentgen Rays.

Assemblyman Reed of the New Jersey State Legislature introduced a bill banning the use of X-rays in opera glasses. And according to Thurston Holland, a London firm advertised X-ray-proof underclothing for ladies. But there were those who feared nothing from the revelations of the rays. In New York, a mysterious 'fashionable woman' who had a 'longing to have a portrait of well-developed ribs' allowed herself to be X-rayed sans stays and corset. The result, termed 'fascinating and coquettish' by the *New York Times*, was exhibited to great public interest.

Doctors and physicists saw the practical potential of X-rays at once, and rushed to experiment with them. Within six weeks of the Vienna *Presse*'s announcement of Roentgen's discovery, the *Electrical Engineer*, an American journal, commented, 'It is safe to say that there is probably no one possessed of a vacuum tube and an induction coil, who has not undertaken to repeat Professor Roentgen's experiments.' So many wanted to repeat the experiments that, by the end of February 1896, Philadelphia and Chicago were sold out of Crooke's tubes (the partially vacuumized glass tubes in which X-rays were produced) and the manufacturers of platino-cyanide were forced to hire extra workers to keep up with demand for fluorescent screens. Within a year of X-rays being discovered, more than one thousand articles and almost fifty books had been published about them.

Most of the early experiments were simply demonstrations of the penetrating properties of X-rays – photographs of the skeletons of living fish, metal objects enclosed in wooden boxes, feet seen through shoes, and so on. It proved much more difficult to get images of the human body that were clear enough to help doctors in diagnosing injuries. One of the first diagnostic X-rays was taken in Montreal on 7 February 1896. John Cox, a professor of physics at McGill University, subjected a young man named Tolson Cunning to a 45-minute X-ray so that doctors could find and remove a bullet from his leg. The following evening, many eager listeners had to be turned away when Cox addressed a crowded public meeting about Roentgen's discovery and his own work. The Cunning X-ray, he noted, 'was clearly under-exposed and should have had at least an hour and a half'. Two days previously, Cox said, he had X-rayed a man with a hip injury, but regretted that 'after one hour's exposure we obtained not a trace upon the plate'. Such disappointments did nothing to decrease the general enthusiasm of the medical profession for X-rays, both as an aid to diagnosis and as evidence in case of future malpractice suits. Many doctors sent patients to be X-rayed in specialized X-ray laboratories. Wolfram Fuchs, an electrical engineer who ran such a lab in Chicago, had performed more than 1,400 X-rays by the end of 1896.

Doctors began to use X-rays not only to diagnose, but also to cure diseases, prescribing 'X-ray séances', as they were sometimes called, for patients who suffered from cancer, tuberculosis, or painful inflammations. No one knew how or why, but sometimes X-ray therapy worked; sometimes it didn't; and sometimes the cure was worse than the disease.

The first systematic practitioner of X-ray therapy was Dr Leopold Freund in Vienna, whose first patient was a five-year-old girl with a hairy mole on her back. In December 1896 she underwent two hours of X-rays every day for 16 days. After 12 days, the hair on her back began to fall out, but her whole back became horribly inflamed and took a very long time to heal. Thereafter Freund limited exposures to ten minutes. 'This accident,' commented the girl's doctor, dryly, 'was full of instruction.'

It soon became apparent that X-rays could cause medical problems, as well as help to solve them. One of the first recorded X-ray injuries occurred in Edison's laboratory in 1896. The inventor was trying to develop an X-ray-powered light bulb. The idea was to produce X-rays inside a calcium tungstate-coated glass tube. The calcium tungstate would light up when bombarded by the X-rays. Clarence Dally, who with his father and three brothers formed a family glassblowing group for Edison, worked on this project after having helped develop the fluoroscope. 'I started in to make a number of these lamps,' Edison later said, 'but I soon found that the X-ray had affected poisonously my assistant, Mr Dally, so that his hair came out and his flesh commenced to ulcerate. I then concluded it would not do, and that it would not be a very popular kind of light, so I dropped it.' Dally continued to work with X-rays through 1898, however, while trying every conceivable remedy for his skin ulcers. Nothing worked. Eventually he underwent many unsuccessful operations to graft skin from his legs on to his damaged hands.

Many X-ray workers were reluctant to believe that the phenomenon that fascinated them and for which they had such high professional and humanitarian hopes could be harmful. In the summer of 1896, a Columbia University student and X-ray enthusiast named Herbert Hawks demonstrated X-ray equipment at department stores and other public places in New York City. Hawks used to focus the rays on his head so that spectators could see his jawbone on a screen. 'At the end of a few days,' he wrote in his diary, 'I found that the rays were having quite an effect upon me.' He became partly bald, and lost his eyebrows and lashes. His fingernails stopped growing; his eyes became blood-shot and his vision impaired. His chest was burnt through his clothing. His hand, which was the most exposed part of his body, swelled and became inflamed. 'The exposure to produce such effects as these probably amounted to between two and three hours.' The doctors Hawks consulted treated him as a case of parboiling.

Though his injuries forced him to stop work after only four days, Hawks went back to the job after being treated. He stopped putting his head into the rays and tried to protect his hand, finally resorting to wrapping it in tinfoil. Hawks's suffering continued for some months after his job ended. Later he wrote an article for the *Electrical Engineer* in which he made light of his injuries, attributing them not to the rays themselves, but to some mysterious electrical effect.

Clarence Dally and those who followed him were victims of the ionizing power of X-rays. Like all radiations, X-rays are a form of energy. They are on the high-energy end of the electromagnetic radiation spectrum, at the low-energy end of which are radio waves, microwaves, and light waves. X-rays do not occur naturally; they are produced when a metal target is bombarded by a stream of fast-moving electrons. They travel with such energy that they can knock electrons out of any atom with which they collide. Normally atoms are electrically stable, with equal numbers of electrons and protons. Losing one or more electrons gives an atom an electrical charge, a process called ionization. Ionized atoms are very unstable and prone to form new combinations with other atoms and molecules. In living tissue, ionization sets off a chain of physical, chemical, and biological changes that can result in serious illness, genetic defects, or death. Scientists do not fully understand the complex interactions that atoms go through in the split-second after they are ionized, but it is clear that these interactions can cause changes in the molecules of human cells. These changes may kill the cells outright, or alter them in such a way that cancer or other physical injuries – such as burns, cataracts, or benign tumours – eventually develop. In theory, even a single 'hit' of ionizing radiation can create irreversible cell damage. There is a consensus that cells are most sensitive to damage when they are dividing, thus foetuses and growing children are especially vulnerable to radiation. Radiation can injure either the person who was exposed to it, or his or her progeny. Somatic injury is injury to the exposed person. Genetic damage is damage to the exposed person's gonads or sex cells, which is passed on to future generations.

The inherent dangers of X-rays were exacerbated in the early days by primitive machines and methods of use. Much early X-ray equipment was home-made, put together by doctors or physicists following descriptions in scientific journals. But home-made or not, early X-ray machines – and the electrical systems that powered them – were extremely unreliable, producing radiation at times too weak to be useful, at other times so strong that it irradiated people in adjoining rooms.

Reminiscing about the first two decades of this century, James Ewing, a pioneeer X-ray therapist, wrote, 'The prescription of dosage was so uncertain and the results apparently so capricious that all one could really do was to place the patient under the machine and hope for the best. Patients were burned from unexpected leaks, and on one or more occasion, it is said, actually electrocuted on the treatment table.'

Though some X-ray subjects suffered horribly from their exposures, the people most at risk were the radiologists and the X-ray tube-makers. They acted as guinea pigs, testing the rays on their hands before X-raying a patient or selling a tube. Each time they X-rayed a patient they themselves were exposed to X-rays also. 'That the radiologist was not more frequently affected by scattered radiation from such exposures was attributable to the fact that during the exposures he might retire to another room to see other patients – or even go out for lunch,' remarked one early user of X-rays. Fluoroscopic examinations were especially dangerous. Since the X-ray was projected onto a screen, but not preserved on film, the exposure had to be long enough to give doctors a chance to thoroughly study the image. Doctors often had to stand in the path of the rays in order to get a good view of the screen.

By the end of 1896, 23 cases of severe X-ray injury, mostly to radiologists and manufacturers of X-ray tubes, had been reported in scientific journals. As the reports of damage came in, many X-ray enthusiasts claimed that electricity, ozone, or faulty equipment, rather than the rays themselves, were to blame. Elihu Thompson, a physicist working at the General Electric Laboratory in Schenectady, New York, attempted in late 1896 to settle the matter by experimenting upon himself. 'I asked myself what part of my body I could best afford to lose and decided it was the last joint on my left little finger.' Accordingly he deliberately held the little finger of his left hand about one and a half inches from the X-ray tube for one half hour to no immediate effect. A week later, however, 'that finger reddened, became extremely sensitive, swollen, stiff and to a certain extent painful . . . There is evidently a point beyond which exposure cannot go without causing serious trouble.' Several weeks later Thompson wrote to a colleague, 'I do not propose to repeat the experiment . . . the whole epidermis is off the back of the finger and off the sides of it also, while the tissue, even under the nail, is whitened and probably dead, ready to be cast off . . . The wound itself is very peculiar and I never saw anything like it. It continued to develop and spread over the extent of the exposed surface for three weeks and I am not sure that the affection has reached its limit.'

Thompson's experiment was widely reported and convinced many radiologists of the importance of taking precautions to avoid inflicting burns on themselves or their patients. A few primitive items of protective equipment made their appearance, including lead shields, filters, and diaphragms to limit the size of the beam. Eventually new X-ray machines were developed that were more reliable and therefore safer for the operator and the patient. Such precautions did much to alleviate the burns that are the immediate hazard of over-exposure. In

1901 there was only one published account of an X-ray burn, a great improvement over previous years. Many radiologists believed that the problem of X-ray injury was solved.

CHAPTER TWO

The First Standards

In 1904, at the age of 39, Edison's assistant, Clarence Dally, became the first person to die as a result of exposure to ionizing radiation. His X-ray burns had developed into cancer. For a time, Dally's doctor had treated him with more X-rays 'in the hope', recorded a consultant physician, 'of undoing what the ray itself had done'. In 1902 after six years of increasing pain from burned, inflamed, and ulcerated skin, Dally submitted to having his left hand, and four fingers and part of the palm of his right hand, amputated. Before his death, Dally's right arm was amputated at the shoulder, his left at the elbow.

Dally was soon followed by a steady stream of 'martyrs to science through the roentgen rays' – to use the title of a book by a radiologist who himself later died of cancer. The infant profession of radiology was shaken. At their annual meeting of 1908, members of the American Roentgen Ray Society listened in silence to a paper describing more than 50 cases of radiation poisoning. The author, Dr Charles Allen Porter, to whom many of the victims had come for skin grafts, told his audience, 'The amount of pain which these patients suffer is variable, though usually extreme. From my experience and personal communications from patients, I believe that the agony of inflamed X-ray lesions is almost unequaled by any other disease.' The *American X-Ray Journal* expressed its concerns symbolically by removing from its cover an allegorical figure of 'Science' holding a Crooke's tube aloft to irradiate the world.

The dangers of over-exposure were clear; the equipment needed for protection against radiation was available, but radiologists continued to injure themselves and their patients. The problem was the lack of precise protection guidelines. No one could say exactly how much protection was required, and doctors were reluctant to expend time, money, and effort on what might be an unnecessary degree of protection. Dr William Rollins, a crusader for careful use of radiation and a brilliant developer of X-ray equipment, tried to solve that problem. Proposing what he hoped would become a routine procedure, Rollins advised X-ray users to expose a photographic plate to their equipment for seven minutes. If the plate did not fog up in that time, the equipment was adequately shielded. If it did, more shielding was necessary. Few people bothered with Rollins's test. They thought it troublesome and arbitrary. It *was* arbitrary: X-ray operators who used Rollins's test received hundreds of times more radiation than is permitted today. Unfortunately, by ignoring Rollins's rule of thumb, they exposed themselves to even more radiation than that.

However common radiation poisoning was shown to be, many radiologists assumed it couldn't happen to them. Dr Charles Leonard of Philadelphia described their attitude: 'A seeming disbelief renders . . . experienced operators careless . . . It may be dangerous for others but not for them. Because they cannot see immediate effects, they cannot appreciate that any injury is being done.'

The disbelievers – who constituted the majority of medical men in the early part of the century – also feared that an obsession with safety would make it impossible for them to carry out their work – and, by impeding the development of radiology, deprive humanity of a wonderful medical tool. 'There is no such thing as absolute protection, unless you go to an adjoining town and operate by telephone . . . I do not believe that it is necessary to cloth ourselves in armor . . . attempts at absolute protection have been carried to rather absurd extremes,' remarked a doctor in Memphis. When Leonard called for all X-ray tubes to be encased in lead shields that would screen out unnecessary, but harmful, secondary rays, many radiologists argued that shielded tubes would be awkward to use. Dr Eugene Caldwell of New York, who already had cancer of the hand, spoke out against the shields and against 'undue alarm over the dangers of the secondary rays'. Caldwell died of cancer in 1918.

Rollins accused radiation enthusiasts of 'attempts to ignore or disparage the crucial experiments that have been reported . . . on the effects of X-light on animals'. Dr Mihran Kassabian, one of the early leaders of radiology, campaigned against the use of the very word 'burn' to describe the effect of over-exposure. Kassabian, according to his biographer, fellow-radiologist Percy Brown, 'was worried that if "burns" came to be associated, in the popular mind, with the medical use of the Roentgen rays, the progress of a new science might be definitely inhibited at a crucial period in its development'. Kassabian died of radiation-induced cancer in 1910; Brown himself died of cancer in 1950.

In 1909 the president of the American Roentgen Ray Society, Dr George Johnson, announced gloomily to his members, 'The insurance companies are beginning to look upon us as undesirable risks.' The insurers were right to be wary. The tragedies of the first years of radiation unfolded slowly. They continued to come to light long after the medical profession's belated efforts to control its powerful new tool.

During the First World War, radiation protection was largely forgotten, but radiation use soared. Both sides relied heavily on portable X-ray machines of a rather primitive design to locate shrapnel and to help in setting broken bones, but the chaos of the battlefield and the inexperience of the operators made safety measures next to impossible. There were many radiation casualties among medics and the men they tended. After the war newspapers took up these cases, drawing public attention to the dangers of radiation.

Nonetheless, doctors extended X-ray therapy to a wide variety of diseases, from birthmarks to syphilis, as more powerful and reliable X-ray machines became available. The editor of the *American X-Ray Journal* said that 'there are about 100 named diseases that yield favorably to X-ray treatment'. Radiation treatment for benign diseases became a medical craze that lasted for 40 or more years. Some of the most valuable information about the biological effects of high doses of radiation comes from studies of large groups of people needlessly irradiated for such minor problems as ringworm and acne. 'Female problems', physical and mental, were thought to respond particularly well to radiation. Many women had their ovaries irradiated as a treatment for depression. One textbook of radiology recommended using radiation to induce menopause; the result, the author promised, 'would be less stormy than that produced by castration'. Excessive bleeding during menstruation was sometimes treated by irradiating the uterus, and doctors could be assured, according to Dr Thomas Cheery of the New York Post-Graduate Medical School and Hospital, that 'sexual desire and response are lessened or lost in only about twenty per cent of cases'.

Entrepreneurs and quacks also saw a feminine use for X-rays. All across the United States, beauty shops installed X-ray equipment to remove their customers' unwanted facial and body hair. The biggest operation was the Tricho Institute, founded by Albert Geyser, a New York physician. Geyser leased X-ray machines to beauty parlours and gave two-week courses to operators. The Tricho System (Geyser and rival promoters discouraged the mention of X-rays; instead they used euphemisms such as 'light treatment', 'Epilax Ray', and 'short-wave treatment') involved high-dose treatments every two weeks, until the hair fell out, and subsequent treatments to prevent it from returning. The effects, as hundreds of doctors across the country soon learned, were appalling. One case, reported by two doctors who studied the women, will give an idea of their sufferings:

> Miss H.K., from Milwaukee, 19 Tricho treatments between July 1926 and September 1927. In October 1927, patient first observed redness and itching, which increased in severity, resulting in painful ulceration and disability. By December 1928, ulcers and other painful changes were present on hands, forearms, and legs, knees . . . Patient wholly disabled for work, bedridden, and suffering severely.

Newspaper accounts of the plight of these women, rather than government action, gradually ended beauty parlour X-ray treatments. In 1947 Doctors A. C. Cipollaro and Max Einhorn reported that, 'As the years passed, cases of radiodermatitis, horrible burns, painful ulcerations and cancer resulting from the Tricho system of treatment were observed by dermatologists in all sections of the country . . . the number of cases of X-ray burns, cancer and death resulting from he treatments administered by the Tricho Institute must have run into the thousands. It is impossible to obtain or estimate the actual number because the cases were not recorded.'

During the 1920s, a second wave of deaths among the pioneer radiologists and their patients shocked the radiation profession. These deaths were caused by blood diseases and cancers with long latency periods. For the first time, radiologists realized that the effects of acute and chronic over-exposure could take decades to reveal themselves. This realization inspired a new round of protection efforts, with medical and radiological societies in many countries publishing safety recommendations and urging their members to observe them. Among the national radiological bodies that took a leading role in radiation protection were the United States Advisory Committee on X-Ray and Radium Protection, and the British X-Ray and Radium Protection Committee.

There were no internationally-agreed safety guidelines until the International Congress of Radiology, a consortium of radiological groups from many countries, adopted some at its second meeting, in Stockholm in 1928. The ICR recommendations were more detailed than any published before. They specified exactly how much lead shielding X-ray tubes of varying voltages should have; how thick the walls of radium storage rooms should be; and the proper size, temperature, and colour scheme (all-black) for X-ray rooms. The safeguards were precise, but arbitrary. Without answers to two key questions, all safeguards were doomed to be arbitrary. The questions were: How much radiation is dangerous? How much protection is enough?

Ever since Roentgen discovered X-rays, scientists had tried to measure X-radiation by its chemical and physical properties – its ability to blacken photographic film, to change the colour of certain chemicals, to cause fluorescence, to ionize gases, and so forth. Doctors preferred systems based on the biological effects of radiation, since they offered the most direct method of predicting how their patients would react to a given dose. One such system employed a 'radiometer' to show when an 'epilation dose' had been administered. An epilation dose is the amount of radiation that would make an average subject's hair fall out. Ordinarily a doctor would have no way of knowing whether he had delivered such a dose, because the hair loss does not usually occur until a few weeks after the exposure. With the radiometer, however, a small capsule of compressed barium platino-cyanide placed halfway between the tube and the subject changed from bright green to orange when the epilation dose was reached, a sign to the doctor to stop.

The most immediate biological effect of radiation is a skin burn known as an erythema, and upon this another system of measurement was based. The amount of radiation required to start turning skin red and inflamed was known as a threshold erythema dose (ED). Larger or smaller amounts of radiation were expressed as fractions or multiples of an erythema dose. For many years the erythema dose was the most popular measure of radiation, but it was an indirect and inexact measure. In the first place, the production of an erythema depends not only upon how much radiation a source emits, but upon many other factors as well, including the length of the exposure, the area exposed, the amount of shielding used, and the time elapsed between exposures. Secondly, individuals vary widely in their sensitivity to radiation. And thirdly, the word erythema meant different things to different people, anything from a relatively mild blistering to a severe inflammation lasting for several months. As a result, the amount of radiation required to cause an erythema varied by as much as 1,000 per cent, depending upon the occasion, the subject, and the observer.

Nonetheless, well into the 1930s the erythema dose was the most common radiation unit used by doctors. Hospitals calibrated their X-ray machines according to it, reports Lauriston Taylor, a physicist and a key figure in the field of radiation protection for more than 60 years. 'They would expose a person, usually the thigh, to radiation at a given distance, given voltage on the tube, given current through the tube, and wait until the skin began to redden.' From this it was possible to calculate roughly the intensity of the radiation each machine emitted, although the early machines – and the electrical systems that powered them – were so variable that the intensity of their output might change from minute to minute.

Physicists looking for more precise ways to measure radiation came up with many techniques. Often these were named after their proposers and shortened to a single initial. Thus, in the first three decades of this century, scientists could measure radiation in B, D, E, e, F, H, Ha, I, K, M, X, and x units. Eventually the Roentgen, honouring the discoverer of X-rays, triumphed as the name for a unit of radiation, but not before several different Roentgen-units, all with different values, but all symbolized by the letter R, had co-existed for many years, adding considerably to the general confusion. Finally in 1928, the International Congress of Radiology's X-ray unit committee, chose the roentgen (giving it a small r as a symbol to distinguish it from the many upper-case Rs previously in use) as the international unit of X-ray measurement. A roentgen was defined as the amount of radiation needed to produce a given number of charged ions in a given amount of air. The adoption of a unit of radiation measurement gave scientists the common language they needed to establish a universal and precise protection standard.

In 1924, Arthur Mutscheller, a physicist employed by a manufacturer of X-ray equipment, decided to try and discover how much radiation humans could tolerate. He began by asking doctors and technicians at several typical X-ray laboratories whether they had suffered any ill effects from working with radiation. Since none had noticed ill effects, Mutscheller concluded that they were receiving tolerable doses. His next job was to measure those doses.

Because there were no instruments sensitive enough, Mutscheller couldn't directly measure how much radiation fell on the workers. The best he could do was to take measurements near the radiation source and roughly calculate what the radiation intensity would be where the workers stood. To do this he used a formula of his own devising – multiplying the electrical current powering the X-ray machine by the length of exposure, and dividing the result by the square of the distance between the subject and the source. To translate this into erythema units Mutscheller then multiplied by 36.8 – a number he chose because he believed it gave the correct result. Taking into account the lead shields in use in the installations he was studying, Mutscheller concluded that workers could be safely exposed to up to 0.01 of an erythema dose every month. In September 1924, he presented his conclusions to the annual meeting of the American Roentgen Ray Society.

When Mutscheller proposed 0.01 of an ED as an acceptable dose, he did not say which hospitals he surveyed, what radiation doses workers at those hospitals were receiving; what protection measures they used, or how big his survey was. Lauriston Taylor, who knew Mutscheller and discussed his proposal with him, reports that he surveyed no more than six hospitals. Mutscheller was well aware of the imprecision of the standard he was proposing. 'It seems,' he wrote, 'that under present conditions and standards accepted at present, it is entirely safe if an operator does not receive every 30 days a dose exceeding 1/100 of an erythema dose . . . This dose, however, is derived from the average of a limited number only of typical examples, and is perhaps not yet sufficiently checked biologically.'

Faith in Mutscheller's findings was strengthened by the fact that several scientists studying the same problem independently reached similar conclusions. In 1925 Rolf Sievert of Sweden estimated that people were exposed to between 0.001 and 0.0001 of an erythema dose from naturally occurring radiation each year. He decided, without conducting experiments or collecting scientific evidence, that humans could tolerate 0.1 of an erythema dose per year without harm – a figure very close to Mutscheller's.

A few years later the British physicists Alfred Barclay and Sydney Cox published their study of two individuals who had worked with radiation for six years without any observable effects. Barclay and Cox figured out how much radiation the two had been exposed to, and divided that by 25, a number they chose completely arbitrarily, to come up with a tolerance dose of 0.08 of an erythema dose per year, roughly equivalent to Mutscheller's and Sievert's estimates.

It was encouraging that three studies arrived at so nearly the same conclusion independently. However, as Taylor has pointed out, the conjunction was simply fortuitous. 'Pure judgement was the only common factor in the three studies contributing to their arrival at the same result.' The findings in all three were essentially educated guesses contingent on many unsubstantiated assumptions, rather than the results of strict scientific observation.

Mutscheller did not at first translate his ED into physical units. In 1926, however, he stated that one erythema dose equalled 1,300 R-units. Unfortunately he didn't specify which of the many R-units then in existence he was using. The variety of R-units, combined with the inherent imprecision of the erythema dose made it very difficult to say exactly how much radiation Mutscheller's standard would allow. In 1926, William Henry Meyer and Otto Glasser found that seven scientists or groups of scientists who translated erythema doses into R-units between 1918 and 1925 had come up with seven different answers ranging from 170 R to 2,500 R. However, there was a mid-range from 1,200 to 1,800 R, which gave some credence to Mutscheller's figure of 1,300 R.

The figures collected by Meyer and Glasser included backscattering – that is, radiation reflected back from the surface of the object being irradiated. To get a figure without backscattering, a German scientist named Kustner in 1927 sent a questionnaire to 12 German X-ray institutions asking how much radiation, not including backscattering, usually produced an erythema in their labs. Seven of the twelve replied to Kustner's questionnaire, giving values between 400 and 650 R. The average of the seven was 550 R, a figure that didn't go very well with Mutscheller's 1,300 R, since it was unlikely that more than half of Mutscheller's reading consisted of backscattering. Nonetheless, 550 R became the accepted figure for one erythema dose without back-scattering. Actually it became common to round the ED up to 600 R, which meant that 0.01 of an ED also went up from 5.5 R to 6 R per month, an increase of about ten per cent.

In 1933 Lauriston Taylor recommended to the US Advisory Committee on X-Ray and Radium Protection, of which he was chairman, that it adopt Mutscheller's tolerance dose, expressed in the new unit, roentgens. Assuming that the erythema dose of 600 R could be translated directly into roentgens, Mutscheller's tolerance dose would be 6 r per month or 0.24 r per day. In March 1934, the US committee adopted Mutscheller's limit, becoming the first organization ever to set a tolerance level for radiation exposure. In the light of the uncertainties surrounding the erythema dose, however, the US committee decided to round the 0.24 r figure down to 0.1 r per day. Using the higher figure, they feared, would, in Taylor's words, 'appear to be implying an unreasonable knowledge of the subject'.

Four months later, the International Congress of Radiology's Committee on X-Rays and Radium Protection also agreed to recommend a tolerance dose based on Mutscheller's work. The international committee also rounded down the 0.24 r figure, but only to 0.2 r per day, thus allowing twice as much radiation as had the US committee. There was no real reason for the difference, according to Lauriston Taylor, who in addition to chairing the US committee, was a member of the international one. 'We thought there was enough conservatism in this all the way along; we just didn't see any difference between 0.2 and 0.1'

The 1934 standards set radiation protection on a new path. Internationally agreed, authoritative, precise, they inspired confidence. Today's protection standards are their direct descendants. But, as a close inspection of their development reveals, those first standards rested on scientifically shaky ground – on studies too short to detect long-term effects; on inadequate samples; on ill defined and inconsistent units of measurement; on untested assumptions. 'Mutscheller's work', says Lauriston Taylor, 'was seriously flawed, and yet that is still the basis for our protection standards of today. It really is.'

The standard-setters knew the shortcomings of the experimental data, but they could not afford to wait for certainty. Physicists, doctors, and the people they treated, desperately needed safety standards. In 1936 the German Roentgen Society erected a monument in Hamburg to the 'martyrs of radiation'. The names of 169 victims were inscribed in stone. Over the next 30 years, as the latent injuries of the early years came to maturity, the memorial was expanded to accommodate several hundred more names.

CHAPTER THREE

Radioactivity

When he heard of Roentgen's discovery of X-rays, Henri Becquerel, a professor of physics at the Ecole Polytechnique in Paris, began to wonder whether luminescent materials – materials that emit light when hit by sunlight – might not also emit X-rays. He placed various luminescent minerals on unexposed photographic plates wrapped in lightproof paper, left them in the sunlight, and then developed the plates to see if the sun had caused the minerals to give off rays that penetrated the paper. On one occasion at the end of February 1896, when he was about to try uranium, the sky was overcast, so he put the plates away, waiting for a sunnier day. On 1 March Becquerel took the plates out of the drawer and decided to develop them, though they had been kept in darkness. To his amazement he saw the outline of the uranium on the plates. He correctly concluded that the uranium was spontaneously emitting a new kind of penetrating radiation. Further experimentation showed that the uranium rays penetrated thin sheets of copper and aluminium, as well as paper. Becquerel made his discovery public the very next evening at the weekly meeting of the French Academy of Sciences. His paper, 'On visible radiations emitted by phosphorescent bodies', was published ten days later.

Marie Curie, looking for a subject for her doctoral dissertation, took up Becquerel's discovery. She published a paper in April 1898, in which she noted that pitchblende, a uranium-bearing ore, emitted more radiation than could be accounted for by its uranium content. She coined the word 'radioactivity' to describe the phenomenon of spontaneously emitted radiations. In December 1898 the Curies identified a new element in pitchblende, radium – evidently a strongly radioactive material. In order to gain official recognition for radium as a new element, it was necessary to isolate enough of it to be seen and measured. Marie Curie's epic struggle to do so is one of the legends of science. Though she and her husband worked together, she bore the brunt of the physical labour. The only space the School of Physics in Paris would give her was an unheated outdoor shed. Her raw material was thousands of pounds of radioactive waste imported from a pitch-blende mine in the Belgian Congo. For years, Marie Curie laboured over buckets and cauldrons, lifting, pouring, and stirring the pitch-blende waste, all the while breathing in its radioactive fumes. So poisoned was the atmosphere that the notebooks the Curies used during this period are still dangerously radioactive. No one knew then the consequences of working, unprotected, with highly radioactive materials. In March 1902, after four years of back-breaking work, she obtained one-tenth of a gram of radium from one ton of radioactive mine waste. In 1934 she died of a form of leukaemia called aplastic anacmia which was most likely induced by her exposure during those years.

For their discoveries, the Curies shared with Becquerel in 1903 the third Nobel Prize in Physics, but it was Ernest Rutherford, a New Zealander transplanted to Montreal and later to England, who discovered what radioactivity really was. Rutherford found that uranium gave off several different types of radiations, which he named alpha and beta. Alpha and beta are high-speed particles that behave much like X-rays and other radiations at the high-energy end of the electromagnetic radiation spectrum. Alpha particles consist of two protons and two neutrons bound together. They have two positive electrical charges and a large mass. Beta particles are just electrons under another name. Each one has a single negative charge and is less than $1/7,000$ the size of an alpha particle. Soon after Rutherford's announcement, Paul Villard, a Frenchman, identified another emission from radioactive materials - gamma rays. Gamma rays are essentially very energetic X-rays, except that instead of being emitted by a machine, they are emitted naturally by certain radioactive substances.

The biological damage these radiations do is a function of their ability to ionize atoms. In ionization, the radiation transfers some or all of its energy to the atom. Alpha particles interact easily with the matter they pass through because they are large and have a strong electrical charge. Thus they travel only a short distance before transferring all their energy. They can penetrate only a few inches in air, and just a few layers of cells in human tissue, which is denser than air. Beta particles can penetrate a few feet of air and the first layers of skin in human beings. Gamma rays and X-rays, having no electrical charge and no mass, do not interact with matter unless they actually collide with an atom. Thus they can penetrate deep into a human body, or even pass straight through one without losing their energy. A thick piece of lead or concrete will stop most gamma rays and X-rays. Theoretically, a single 'hit' of radiation to an atom can destroy the cell that the atom is part of, but a cell's chances of being destroyed obviously increase as more of its atoms are ionized. Thus, because its ionization is concentrated in a smaller area, a given amount of alpha radiation causes more biological damage than the same amount of beta rays, gamma rays, or X-rays.

Rutherford also described the structure of the atom. He found that each atom is like a tiny solar system – a small, heavy nucleus surrounded by orbiting electrons. The nucleus consists of a tightly bound collection of protons and neutrons. There are always the same number of electrons as there are protons, but the number of neutrons may vary. The number of protons in an atom determines what element it is. If two atoms have the same number of protons, but different numbers of neutrons, they are two different 'isotopes' of the same element. Isotopes of a single element are identified by their 'mass number', which is the total number of protons and neutrons they contain. Most elements are mixtures of between two and ten isotopes. Naturally occurring uranium, for example, is a mixture of three isotopes, U–234, U–235, and U–238.

Some atoms, called radionuclides, are naturally unstable. Their nuclei are constantly degenerating – giving off alpha, beta, or gamma radiation until they have achieved a stable state. As they emit radiation, they change, or 'decay', into different isotopes. Uranium–235, for example, goes through fourteen changes before finally becoming a stable isotope of lead. Every radioactive isotope has its own 'half-life', which is the time required for half of the atoms of the isotope to decay to a different form. Radium–226, for example, has a half-life of 1,622 years. In that time, half of any amount of it will have decayed to radon–222, which itself has a half-life of 3.8 days before it decays to polonium–218. The more unstable an element, the faster it decays. Polonium–214 has a half-life of only 0.00016 seconds. Uranium–238 is much more stable; its half-life is 4.5 billion years.

Becquerel's discovery of radioactivity did not attract much popular attention, but radium was a different matter. Not only was it an entirely new element, and a highly radioactive one, but it had been discovered by a woman. Radium quickly became a popular obsession. The American chemist Henry Bolton suggested that bicycles would 'be lighted with disks of radium in tiny lanterns'; a farmer proposed mixing radium with chicken-feed so that hens would lay hard-boiled eggs; and the dean of Columbia University's College of Pharmacy, Dr H. H. Rusby, said radium-fertilized soil would produce more and better-tasting crops.

One of the first to try and put radium to some practical use was Loie Fuller, the American 'light fairy' whose dance using electric lights and floating veils had made her the star of the Folies-Bergère. Fuller wrote to the Curies, asking if radium could be used to make a luminescent costume. Their answer, though disappointing, was so kind that Fuller asked to be allowed to dance privately for them in their home. Somewhat bemused, the Curies agreed. Arriving at their house in the Boulevard Kellermann with a troop of electricians one day, the dancer spent several hours re-arranging the furniture in the narrow dining room and setting up the electric lights for her dance. The evening was such a success that it was repeated several times. Marie Curie was charmed by what her daughter Eva called the dancer's 'delicate soul', and the unworldly scientists and the music hall star became friends.

In April 1901 Becquerel borrowed from the Curies a tube containing a small amount of a radium derivative. After carrying the tube in his waistcoat pocket for only six hours, Becquerel discovered that it had burned his skin through several layers of clothes. The doctor he consulted diagnosed the injury as identical to an X-ray burn. If radium had the destructive power of X-rays, Becquerel and others reasoned, it might also possess their therapeutic power.

Becquerel was not the first to recognize the potential for healing and harm of radioactive substances. Soon after a Berlin apothecary named Martin Klaproth discovered uranium in 1789, chemists realized that the salts of uranium made excellent dyes, and the mining of the raw ore, pitchblende, became a boom industry. It was also found that uranium was a potent poison if injected intravenously, and that it caused kidney disease. In the late nineteenth and early twentieth centuries, uranium salts and remedies such as 'uranium wine' were used in the treatment of diabetes, stomach ulcers, and consumption. Results were disappointing, however, and uranium preparations disappeared from pharmacists' shelves some fifty years ago.

The gamma rays that caused Becquerel's burn were more penetrating, and thus potentially more hazardous, than the early X-rays (which were less energetic, or 'softer', than gamma rays). On the other hand, radium had the advantages of being easily portable and emitting a steady intensity of radiation. In 1903, Alexander Graham Bell envisioned its medical use: 'There is no reason why a tiny fragment of radium sealed up in a fine glass tube should not be inserted into the very heart of a cancer, thus acting directly upon the diseased material.' Until about 1910 radium was in very short supply and prohibitively expensive. However, with the discovery of large uranium deposits in Canada and Colorado, radium became a popular treatment, especially for cancer. At $120,000 per gram, the price was still staggering, but a fraction of a gram was enough for most medical uses.

Well into the twentieth century, many doctors believed radium to be totally safe. In 1916 the medical journal *Radium* asserted that 'Radium has absolutely no toxic effects, it being accepted as harmoniously by the human system as is sunlight by the plant.' Doctors learned about radium the same way they did about X-rays: by trial and error. Speaking in 1953 at the American Radium Society's annual dinner, Dr James Case, a key figure in the development of radiation therapy, recalled, 'At first we gave very large doses . . . so large in fact, and within so short a time, that most of the patients were forced to take to their beds for a week or so, and some of them were really prostrated. We often gave blood transfusions, and other supportive treatment.' The conditions for which radium was used ranged from heart trouble, to impotence, to ulcers. Some doctors prescribed injections of radium as a tonic for patients who were depressed or 'under the weather'.

In the early days doctors simply strapped flat containers of radium on to the surface of whatever part of the body they wanted to treat. To reach deeper injuries, they injected radium chloride intravenously or slipped small capsules into natural body cavities. That technique often resulted in severe burns – for doctor and patient. Before the First World War, Dr John Hall-Edwards, a pioneer radiologist from Birmingham, England, treated a young girl who was suffering from tuberculosis of the skin with a radium source that he had wedged into the end of a bamboo pole. Andie Clerk, who witnessed the operation as a friend of the girl's family, recalled many years later that 'He touched the skin with it very carefully, but it wasn't effective, rather defective. It made her slightly out of her mind, though I believe she got that back in later years. He was handling something which he knew was dangerous but didn't know to what extent. A few years later he lost both his hands as a result of contact with it.' Hall-Edwards said of his radiation injuries that 'the pain experienced cannot be expressed in words'.

Eventually, more efficient applications were developed: small amounts of radon, a radioactive gas that is a decay product of radium, was sealed into tiny gold, platinum, or glass capsules which could be inserted directly into the diseased tissue, reducing the irradiation of healthy tissue. As the means of application improved, and doctors gained confidence in the new techniques, the use of radium in treating cancers grew. In 1915 and 1916, 424 malignant tumours were treated with radium at one hospital alone, Memorial Hospital in New York. Of these, 120 disappeared completely, at least for a time.

The long-term effects of radium therapy were as yet unknown, but the history of medical X-rays was reason enough for caution. One ill omen was that most of the doctors using radium had no training in doing so. In 1931 the US Bureau of Mines, which was the chief source of radium in the United States, ascertained that 287 hospitals and 414 individual physicians had supplies of radium. Yet, there were then fewer than 50 physicians in the entire country who specialized in radiation therapy, according to Dr Juan del Regato, founder of the first centre for radiotherapy training and historian of the American College of Radiology. Despite his high hopes for radium therapy, Dr Henry Janeway, the head of Memorial Hospital's radium therapy division, warned his fellow practitioners in 1917 of the need to proceed gingerly. 'The disastrous effects of over-exposure are so serious, and they inflict on patients already pitiable so much additional suffering, that too great care cannot be taken to avoid it. It takes at least two or three months to know the full consequences of some of these more deeply penetrating exposures, and before the operator is aware of it, he will be deeply regretting the fact that instead of relieving suffering he has increased it.'

On the fringes of the medical profession, a brisk trade in radium-based patent medicines developed, thriving well into the 1930s. Radium preparations were touted for the cure of nearly every imaginable disease from arthritis to cancer to high blood-pressure to blindness. Whatever its medical potency, radium certainly had the power to inspire charlatans. Among the products they successfully peddled were radioactive belts for wearing around any part of the body that needed healing; the *Radium Ear*, a hearing aid with the magic ingredient,

'Hearium'; radioactive toothpastes for cleaner teeth and better diges-
tion; face creams to lighten the skin; and many more products of
dubious safety and efficacy. In 1932, Frederick Godfrey, a self-
proclaimed 'well-known British Hair Specialist', was advertising 'one
of the most important scientific achievements of recent times', a
radioactive hair tonic. A radium-laced chocolate bar was sold in
Germany as a 'rejuvenator'. As late as 1953 a company in Denver was
promoting a radium-based contraceptive jelly.

One of the most popular radium preparations was radium water, sold
as a general tonic and often referred to as 'liquid sunshine'. One
company in New York claimed to supply radium water to 150,000
customers. Thousands more households made their own radium tonic
by filtering ordinary tap water through radium-impregnated crocks.
Many people were defrauded, paying high prices for radium waters
that contained no radium at all, including the well known brand, Radol.
They were the lucky ones. The unlucky ones got what they paid for.
One brand of radium water, *Radithor*, was so radioactive that several
faithful users died from radium poisoning, including the Pittsburgh
industrialist and national amateur golf champion, Eben Byers, who
drank two two-ounce bottles every day for several years. At first Byers
felt rejuvenated and, convinced that he had found the Fountain of
Youth, sent cases of radium water to his friends. But eventually he fell
ill and, in 1932, died painfully in the throes of anaemia, a brain abscess,
and the decay of the bones in his jaws.

The radium water craze also gave a boost to spas with radioactive
waters, some of which, like Hot Springs, Arkansas, now a national park,
still flourish. In 1952, an article in *Life* magazine about the alleged
health benefits of radon sent thousands of arthritis sufferers into
abandoned mines to breathe in 'nature's own remedy'. The Merry
Widow Health Mine, near Butte, Montana, and its more reassuringly
named neighbour, the Sunshine Radon Health Mine, still advertise for
'sufferers of arthritis, sinusitis, migraine, eczema, asthma, hay fever,
psoriasis, allergies, diabetes, and other ailments' to spend an hour and a
half in the mines three times a day for ten or eleven days. 'People who
stay less than eleven days,' warn the owners of the Merry Widow in a
1985 brochure, 'usually later tell us that they received benefit but know
now that they would have benefited more and the benefit would have
lasted longer, if they had made more visits to the mine.'

CHAPTER FOUR

The Dial Painters

'The time will doubtless come when you will have in your own house a room lighted entirely by radium . . . The light, thrown off by radium paint on walls and ceiling, would in color and tone be like soft moonlight.' Dr Sabin von Sochocky made this prediction in 1921, six years after he developed a radium-based luminous paint with the brand-name 'Undark'. Sochocky was also an enthusiastic amateur artist who mixed his own radium oil-paints. 'Pictures painted with radium look like any other pictures in the daytime,' he reported, 'but at night they illuminate themselves and create an interesting and weirdly artistic effect.'

Like other commercial radium paints, Undark's precise formula was a closely guarded secret, but its basic ingredients were radium–226 and luminescent zinc sulphide. In 1915 Sochocky and some partners founded the Radium Luminous Materials Company (later renamed the US Radium Corporation) to market Undark products. The plant, in West Orange, New Jersey, was only two blocks from Thomas Edison's famous laboratory – the one in which he had tried to X-ray the living brain for William Randolph Hearst; in which he had worked on the abortive X-ray light; and in which his assistant Clarence Dally was exposed to the radiation that killed him.

Despite Sochocky's romantic notions about radium, the US Radium Corporation was in the prosaic business of painting numerals on wristwatch dials, with a sideline in glow-in-the-dark crucifixes and light pulls. The company employed mostly women and girls – some as young as twelve. At peak periods as many as 250 workers sat at long rows of workbenches in a second-floor room, whose large north-facing windows flooded the room with light. The dial painters were paid by the piece, so speed counted, but they enlivened their work with jokes and high jinks – some times painting their teeth or fingernails to glow in the dark for special occasions.

With the United States' entry into the world war in 1917, the US

Radium Corporation became an important supplier of luminous instrument dials for airplanes and other military gear. One in six American soldiers wore a radium-lighted watch. To keep up with the demand, each woman had to paint hundreds of dial faces a day – a delicate job, as described by Robley Evans, a physicist who later studied the workers: 'In painting the numerals on a fine watch, for example, an effort to duplicate the shaded script numeral of a professional penman was made. The 2, 3, 6 and 8 were hardest to make correctly, for the fine lines which contrast with the heavy strokes in these numerals were usually too broad, even with the use of the finest, clipped brushes. To rectify these too broad parts, the brush was cleaned and then drawn along the line like an eraser to remove the excess paint. For wiping and tipping the brush the workers found that either a cloth or their fingers were too harsh, but by wiping the brush clean between their lips the proper erasing point could be obtained.' Each time they licked their paint-laden brushes to a point, the dial painters swallowed an infinitesimal amount of radium.

The radium paint business continued to flourish after the war. In 1920 the United States produced four million radium-lighted watches, along with radium-lighted fish bait, doll eyes, gun-sights, and stick-on 'locator buttons' that could be applied to bedposts, slippers, or glasses of water on the nightstand. There were perhaps 50 radium paint studios, employing more than 2,000 dial painters, and the US Radium Corporation was one of the largest.

By 1924, however, a shadow hung over the company. In three years, nine of the young dial painters had died – apparently from a variety of unrelated natural causes. The death certificates, signed by family doctors, cited many different causes of death, including stomach ulcer, syphilis, trench mouth, phosphorus poisoning, necrosis of the jaw, and anaemia. In addition, many of the living dial painters were seeing dentists for severe problems with their teeth and jaws.

Although the destructive power of gamma emissions from radium was well known, radium paint was not thought to be dangerous. First of all, it contained only a tiny proportion of radioactive materials: one part of radium to 30,000 or more parts of zinc sulphide. Secondly, although radium emits alpha particles, which are more biologically damaging than gamma rays, alpha particles were not considered a significant danger to health because they cannot penetrate skin, though they can cause skin cancer or clouding of the cornea, if they get on the surface of the skin or in the eye. Thirdly, whatever radium the women did swallow would have been expelled from their bodies almost immediately – or so it was assumed.

In early 1924 the local Board of Health asked the Consumers' League of New Jersey, a voluntary group concerned with employment of women and children, to examine working conditions at US Radium. Katherine Wiley, the group's secretary, noted that four of the dead women had undergone repeated operations on their jaws, and that eight still-living former dial painters were desperately ill with similar problems. But she found working conditions in the factory to be very good, and the New Jersey State Department of Labor, which also examined the plant, agreed. US Radium Corporation officials assured Miss Wiley that radium was not toxic. They maintained that the cause of the troubles was poor dental hygiene on the part of the victims.

In September 1924, a well-known New York dentist and doctor, Dr Theodore Blum, to whom one of the dial painters had been referred by her family dentist, published an article on oral surgery in the *Journal of the American Dental Association*. A footnote to the article marked a turning point in the case of the dial painters. In the footnote Blum mentioned that in the fall of 1923 he had observed a case of infection of the jawbone 'caused by some radioactive substance used in the manu-facturing of luminous dials for watches'. As Blum later explained, 'I knew I had never seen what she had before. Clinically I couldn't diagnose a thing, but she told me where she worked, and I surmised that her jaw had been invaded – yes, and pervaded – by radioactivity. And so I made my suggestion.' Blum's footnote did not make head-lines, but it caught the eye of Dr Harrison Martland, medical examiner for Essex County, home of the US Radium factory. Martland knew Blum by reputation and was impressed by his suggestion. He began studying the problem and determined to perform autopsies on the next US Radium Corporation workers to die.

Meanwhile, Wiley, unaware of Blum's and Martland's work, and not convinced by the company's arguments, went to New York to consult Florence Kelley, a passionate and effective crusader for the rights of women and children, who was then head of the National Consumers' League. They decided to ask Dr Frederick Hoffman, the Prudential Life Insurance Company's statistician, to make his own investigation. He agreed, and in May 1925 reported his findings to the annual meeting of the American Medical Association. From a statistical point of view, Hoffman said, the number of deaths and illnesses involving anaemia and infected mouths among former employees of US Radium could not be coincidence. 'We are here dealing with an entirely new occupational affection,' probably the result of radium poisoning, he said, which should be covered by the government's industrial diseases compensation scheme. Hoffman reached his conclusions with the help of Sochocky, who – troubled by what was happening to the dial painters – had resigned as technical director of the company he helped found and had offered his help to those investigating the deaths.

Hoffman also told the meeting of his correspondence with the US Radium Corporation. 'It had, of course,' he said, 'become aware of the insinuations made from time to time that work in the plant was injurious to health, but . . . I was informed that technically the opinion seemed to be that the minute quantity of radium introduced into the mouth could not possibly have caused the amount of damage elsewhere indicated.'

What neither Hoffman nor any of the other people studying the health of the dial painters knew was that in March 1924 – more than a year before Hoffman's speech, the US Radium Corporation had secretly asked Cecil Drinker and two colleagues from the Harvard School of Public Health to report on working conditions at the plant. In commissioning the study, the company's president gave his opinion that 'we are suffering from a hysterical condition brought about by coincidence'. In subsequent correspondence he asserted that 'radium in small doses is a stimulant'. Nonetheless, before commissioning the Harvard report, the company had instructed supervisors to tell the workers not to lick their brushes.

On their first visit to the US Radium plant, the Harvard team found the work area spattered with radium paint. Examining workers in a dark room, they found that 'their hair, faces, hands, arms, necks, the dresses, the underclothes, even the corsets of the dial painters were luminous'. Dust collected from the light fixtures and wall beams of the painting room and from offices in the plant glowed in the dark. Gamma rays emitted by the radium fogged up sealed dental films placed in the painting room within two or three days – at least five times faster than considered acceptable. Tests of 22 employees failed to find a single one whose blood count was normal. Reporting to the company in June 1924, Drinker and his colleagues stated that all the workers were exposed to excessive radiation, in several ways: externally from gamma rays, and internally by alpha particles swallowed and inhaled. 'It seems necessary, therefore,' they wrote, 'to consider that the cases described have been due to radium.'

The company was displeased with the Drinker report, and blocked its publication with the threat of a lawsuit. When Drinker learned of Hoffman's scheduled address to the AMA on 'Radium Necrosis', he begged US Radium to publish his report, arguing that such a step would show that the company was working to solve the problem. Company officials turned Drinker down, but they secretly gave the New Jersey Department of Labor a heavily edited version of his report which appeared to absolve US Radium of all responsibility for the deaths.

At about the time of Hoffman's speech, Harrison Martland obtained the first positive proof of radium poisoning. In May 1925 he examined two women, both dial painters – who were suffering from 'extensive jaw necrosis and severe anemia'. Both died soon thereafter, and Martland was able to conduct autopsies, the first done on any US Radium Corporation workers. Working with Sochocky, Martland measured high levels of radioactivity in the women's bones and organs. He also tested a number of living dial painters and found that their bodies contained so much radioactive material that when they exhaled on to a zinc sulphide screen, it glowed.

In their investigations, Martland and his associates learned a great deal about the behaviour of ionizing radiation inside the body. They discovered that when radioactive materials are eaten or inhaled, they do not pass straight through the body, as had been thought. Instead they accumulate in various organs, continually irradiating the surrounding cells. Like calcium, which has a similar chemical structure, radium tends to concentrate in bones. There it can cause bone cancer and damage bone marrow, the tissue in which blood cells are formed.

Martland's findings were also significant for the then-popular medical practice of treating various diseases with intravenous injections or oral treatments of radioactive substances. Thousands of patients were injected with or fed radium for every conceivable disease, including rheumatism, high blood-pressure, menstrual irregularities, depression, waning sex-drive, and something called 'debutante's fatigue'. In 1931, for instance, 31 patients at Elgin State Hospital in Illinois were injected with radium as a treatment for schizophrenia. This practice was sanctioned by the American Medical Association from 1914 until 1932. Martland was the first to point out the inherent danger of such uses of radium. 'Even when soluble bromides or chlorides of radium or mesothorium [an early name for radium–228] are used,' he warned, 'they are precipitated in the blood stream . . . and may accumulate in the organs.'

One of Martland's oddest findings was that the appearance of excellent health may be an early symptom of radiation poisoning. Initially the body defends itself against an invasion of radium by producing many more red blood-cells than normal. The victim looks and feels especially healthy – for a while. But the body cannot maintain this defensive effort forever; sooner or later the radiation-damaged cells take over.

The 'false health' factor and the natural time-lag of radiation-induced diseases make it difficult to diagnose radium poisoning in its early stages, Martland pointed out. He gave the example of a US Radium worker whom he examined in 1925 as part of his dial painter survey. She was 'apparently in perfect health', but tests showed her to have accumulated so much radium in her body that she was radioactive; her breath lit up a fluorescent screen. Within three years she was, Martland wrote, 'a victim of chronic bone lesions of a crippling nature'.

In the spring of 1925, Edwin Lehman, the US Radium Corporation's chemist, was also in apparent good health. Within a month he was dead of acute anaemia. An autopsy found radioactive materials in Lehman's lungs, as well as in his bones and other organs. Though Lehman had not swallowed any paint, he had regularly inhaled radium-contaminated dust, radon gas and other decay products of radium. Radium was then selling for more than $3,000,000 an ounce; at that price Lehman's body contained less than 70 cents worth, an infinitesimal amount, but one that made his bones so radioactive that, left on an unexposed photographic plate, they photographed themselves. Lehman's radioactive body was the first indication that humans could become as contaminated by inhaling radioactive materials as by swallowing them. In his report of the case, which was held up for more than a year by lawyers working for the US Radium Corporation, Martland concluded that 'complete protective measures must be installed in all manufacturing plants, laboratories, hospitals and private offices in which radioactive substances are handled'.

Martland's findings did not alter US Radium's position that poor personal hygiene, not radium poisoning, was killing the dial painters. The company sought scientific backing first from the Harvard team, and, when that backfired, from a new consultant, Dr Frederick Flinn of Columbia University, a specialist in industrial hygiene. Flinn obliged by reporting in December of 1926 that 'an industrial hazard does not exist in the painting of luminous dials'. Flinn acknowledged that the women ingested radium while pointing their brushes with their lips, but said that 98 per cent of what they swallowed was excreted within a few days. He suggested that the deaths were due to a bacterial infection of the jaw. Workers at other factories in the USA and abroad also licked their brushes, said Flinn, yet none of their jawbones had decayed. Actually, studies had shown that luminous watchmakers in France and Switzerland worked differently and did not lick their brushes.

Six months before publication of his article absolving the US Radium Corporation, Flinn examined a woman who had painted dials for another company in Connecticut for fourteen months in 1921 and 1922. By 1924 her bones were so rotted that her leg broke as she was simply walking along. X-rays showed severe bone decay. Her breath was radioactive, and her body was giving off measurable amounts of gamma radiation. The patient also had a profound shortage of red blood-cells. She died in January 1927. In 1928, after one of her workmates also died, Flinn moderated his opinion: 'These two cases, appearing in another state from the first cases, have caused me to suspect that radioactive material is at the bottom of the trouble, even if the mechanism by which it is caused is not altogether clear and not previously suspected.'

Martland's reaction to Flinn's retraction was brief and bitter. 'It is worthy of note that not only were these cases suspected by Hoffman, Castle and the Drinkers in 1925, two and a half years before these statements by Flinn, but the disease was accurately described in all its features by me and my associates two years before Flinn finally made the foregoing admissions.'

Untangling the mystery of radium poisoning did not stop the disease. By 1928 at least 15 women in New Jersey and several in Connecticut, had been identified as victims of radium poisoning. There was no telling how many more radium deaths had been misattributed to trench mouth, anaemia, or syphilis. One family who had been told that their young daughter died of syphilis, exhumed her body five years later, and must have been almost relieved to discover from her radioactive bones that it was radium that killed her. Misdiagnosis remained a problem even after the dangers of radium paint were recognized. 'The late development of symptoms, often occurring from one to seven years after patients leave the employment, and their resemblance to various other diseases would make a diagnosis almost impossible if the patients had moved to another locality,' wrote Martland. 'Physicians not aware of this condition might treat a victim for sepsis, anemia, Vincent's angina, rheumatism, or "God knows what" . . . There is no way of ascertaining in these cases how many have lost their lives and how many have been permanently or temporarily harmed.'

On 10 March 1925, even before Martland had carried out the first autopsy, a 12-line story, headlined 'Girl Says Radium Poisoned Her', appeared in the *New York Times*. Margaret Carlough, 24, a dial painter at the West Orange plant, had filed a $75,000 lawsuit against US Radium Corporation. 'Mrs Carlough,' the *Times* said, 'alleged that she was compelled to wet the tip of her brush with her tongue.' From the announcement of this first lawsuit, the US Radium Corporation was rarely out of the courts or out of the news.

The families of several of the dead dial painters sued the company for damages, as did Dr Lehman's widow. But the case that got the most attention was filed in May 1927 by five former US Radium dial painters – 'The Five Women Doomed to Die', as the newspapers called them. All five had painful and crippling bone diseases. One was so weakened that her thigh fractured spontaneously. Another, after three years as a dial painter, had undergone 20 operations on her jaw and then developed discharging ulcers under her chin; damage to her spinal cord had also paralysed her legs. Two of the five had severe anaemia. 'Newspapers took these five dying women to their ample bosoms,' reported *Time* magazine. 'Heartbreaking were the tales of their torture.'

The expense of dying this way was considerable and because radium poisoning was not covered by New Jersey's workmen's compensation law, the women each sought damages of $250,000 from their former employer. They were so wasted that they had to be assisted to the witness stand; two had actually to be carried to the front of the courtroom. One of the women was unable to raise her right hand to take the oath. The company maintained that there was no scientific proof that the dial painters' injuries were caused by radium. Its lawyers, however, chose not to fight on that issue. Instead, they argued that the women had no right to sue because New Jersey's statute of limitations required injury claims to be filed within two years of the inception of the disease. The newspapers cried foul because the diseases were unde-tectable until several years after their inception, but the court accepted this argument.

The women petitioned against this ruling, but legal manoeuvring held the case up for more than a year, during which time the five, facing certain and painful deaths, became a *cause célèbre*, the focus of well-meaning but distressing publicity. Quacks and faith-healers from around the world sent them nostrums, prayers, and advice. On the presidential campaign trail in 1928, socialist candidate Norman Thomas paid tribute to the dial painters and criticized the US Radium Corporation, its lawyers, and its insurance agents for 'cheating them of this compensation'. Even Marie Curie chimed in with a suggestion that the five eat raw calf's liver to combat their anaemia. Curie herself had only a few years to live before the anaemia caused by her own prolonged exposures to radium proved fatal.

On 10 May 1928 the *New York World*, whose chief editorial writer, Walter Lippmann, was an admirer of Florence Kelley, reviewed the case in an editorial, that concluded, 'We have set down the facts as soberly and as coolly as we know how to do it. Having done that, we confidently assert that this is one of the most damnable travesties of justice that has ever come to our attention.' Thirteen days later, the women were granted permission to sue in the New Jersey Supreme Court. US Radium, however, was still determined to deny legal responsibility for their injuries. The case promised to drag on for years, when suddenly William Clark, a federal judge unconnected with the case, offered to negotiate an out-of-court settlement. Within five days terms were agreed. The company, 'actuated solely by humanitarian considerations', its lawyers insisted, gave each woman a lump sum of $10,000 and a pension of $600 per year, plus medical expenses, for as long as – in the view of a committee of three experts – they suffered from radium poisoning.

Less than six months after the case was settled, Sabin von Sochocky died. His bone marrow, destroyed by radiation, had stopped producing blood cells. The radium inside him had also eaten away at his hands, his mouth, and his jaw. In 1925 while helping Martland test the dial painters, Sochocky discovered that his own breath registered more radioactivity than any of the living dial painters. He knew then what his fate was. In a moving tribute to Sochocky, Martland wrote, 'He died a horrible death. . . During the time I knew him he gave all that was in him to help and comfort others suffering from this disease. Without his valuable aid and suggestions we would have been greatly handicapped in our investigation.'

CHAPTER FIVE

Setting Limits on Radium

By their deaths and terrible injuries, the dial painters taught scientists and doctors that internal exposure was dangerous and should be controlled. In 1928 Florence Kelley successfully agitated for the US Public Health Service to take up the problem of radium poisoning. The result was a series of vague recommendations – that factories install adequate washing facilities and give workers routine medical exams. After these mild safety improvements, the pressure for further controls waned. Work to find out how much radium the human body can tolerate went on, but it went on slowly. Though the scandal of the radium poisonings showed that radium intake should be kept below a certain level, no one had any idea what that level was.

The only standards for radium in those days related to external exposure from gamma and beta rays – and even they were ambiguous. In 1928, the International Committee on X-Ray and Radium Protection could only recommend that radium be well shielded, handled with forceps, and stored in a lead safe. There was no indication from the committee of what level of radium exposure could be tolerated. In 1934, the US Advisory Committee on X-Ray and Radium Protection said that its X-ray limit of 0.1 r per day 'may be used as a guide in radium protection', but cautioned that 'the calculation of radium dosages is not easy, and too great reliance is not to be put on the above figures'. The US group also advised people working with radium to take at least six weeks' vacation a year, and to get as much fresh air as possible all year round. This advice, apparently intended to improve the general health of radium workers, naturally had the effect of attracting people to work with radium.

But none of these guidelines concerned internal exposure to alpha particles, which are mainly dangerous once they are inside the human body. A new limit was needed for internal exposure to radionuclides that are swallowed, inhaled, or injected. In the 1930s, the Massachusetts Institute of Technology was the main centre for radium exposure studies in the United States. Between 1936 and 1938, Robley Evans, now emeritus professor of physics at MIT, and others fed radium to rats in response to a request from the Food and Drug Administration, which wanted to know how much radium it should allow in face cream, contraceptive vaginal jelly, and other radium-containing consumer products then on the market. Unfortunately, recalls Evans, 'we found that these rats were several hundred times more resistive to skeletally retained radium ... than the few human cases which we had studied. We concluded that permissible levels for man had to be determined from observations on man.' This would take time but, as Evans later said, 'There was no pressure [to come up with a tolerance limit] until World War II.'

By early 1941, Evans and his colleagues at MIT, in conjunction with Harrison Martland in New Jersey, had studied 27 people who had been exposed to radium, through dial painting or medical treatments. The US was then gearing up for war, and placing large orders for luminous instrument dials. As Evans recalls it, 'US Navy Medical Corps Captain Dr Charles Stephenson took the lead in insisting that standards of radiation safety be set for the radium-dial industry. Captain Stephenson told me that I must soon provide him with safety standards or else he would have me inducted into the Navy and force me to do it.' As a result of Stephenson's prodding, the National Bureau of Standards set up a nine-man committee, consisting of Leon Curtiss of the bureau, representatives of industry and government, and four people who had studied radium – Harrison Martland, Frederick Flinn, Robley Evans, and Gioacchino Failla, a physicist at New York's Memorial Hospital.

The committee met only once, on 26 February 1941. Curtiss presented a draft text, setting out rules for inspection, ventilation, housekeeping, medical exams, and other non-controversial matters. According to Evans, 'The committee quickly agreed on the housekeeping, etc., rules, and settled down to fill in the blank spaces Curtiss had left for the tolerance level.' The committee reviewed the 27 MIT cases. Those who had less than half a microcurie of radium in their bodies had no detectable injuries. (The curie is a measure of radioactivity for all radioactive substances; one curie is defined as the amount of radioactivity produced by one gram of radium. One microcurie is one millionth of a curie.) Injuries *were* detected in all those whose bodies contained more than 1.2 microcuries of radium. There really was not enough evidence to set a firm limit, but Evans proposed to his fellow committee members that they make an '"informal judgement" decision', setting the limit at 'such a level that we would feel perfectly comfortable if our own wife or daughter were the subject. I then asked each committee-man individually in turn if he would be contented with 0.1 microcurie. Unanimously, they all were.'

The committee recommended that any worker whose body contains more than 0.1 microcuries of radium, as revealed by the amount of radon in his breath, 'change his occupation immediately'. In a related recommendation, it limited the amount of radon that could be in the air breathed by workers to ten trillionths of a curie (10 picocuries) of radon per litre of air. Lastly, the committee extended the 0.1 r per day X-ray limit to gamma ray exposure.

The results of the 26 February meeting were issued in a National Bureau of Standards handbook on 2 May, seven months before the bombing of Pearl Harbor and almost a year after the then-secret discovery of plutonium, a man-made alpha-emitting radioactive element that was to be the core of the Manhattan Project. No one knew what effects plutonium would have upon humans, and there would be no time to find out before it went into large-scale use. Radium, therefore, became the benchmark for scientists studying the health effects of plutonium and other new radioactive materials crucial to the development of the atomic bomb. By studying the effects of plutonium and radium upon animals, scientists on the Manhattan Project were able to compare the toxicity of the two elements, and to derive from the radium limit a limit on human exposure to plutonium.

As Merril Eisenbud, a Manhattan Project veteran, remarked to writer Daniel Lang in the 1950s, 'Historically speaking, the New Jersey cases and the others, coming when they did, were a most valuable accident . . . If it hadn't been for those dial painters, the [Manhattan] project's management could have reasonably rejected the extreme precautions that were urged on it – the remote-control gadgetry, the dust-dispersal systems, the filtering of exhaust air – and thousands of Manhattan District workers might well have been, and might still be, in great danger.' The painful deaths of the early victims of radium poisoning led to better working conditions for the wartime makers of the atomic bomb.

PART TWO

CHAPTER SIX

Building the Bomb

One day in September 1933, the Hungarian physicist, Leo Szilard, was walking near the British Museum, pondering something that Ernest Rutherford had said at a recent meeting of the British Association for the Advancement of Science. Rutherford, then the dean of British physicists, had labelled as 'moonshine' the idea that man could use the tremendous energy locked inside the atom. As Szilard waited for a traffic light to change, he realized, all at once, how to make an atomic bomb.

'It suddenly occurred to me,' he later recalled, 'that if we could find an element which is split by neutrons and which would emit two neutrons when it absorbed one neutron, such an element, if assembled in sufficiently large mass, could sustain a nuclear chain reaction. I didn't see at the moment just how one would go about finding such an element, or what experiments might be needed, but the idea never left me ... in fact became a sort of obsession with me.' Szilard had identified the two conditions essential to producing an atomic explosion – critical mass and chain reaction. His intuition was years ahead of its time.

Szilard told a number of prominent scientists about his idea, but he was not a nuclear physicist, and, in his words, 'I couldn't evoke any enthusiasm.' Frightened of their potential for harm, Szilard kept his theories secret and filed for British patents on them. His motive was not personal wealth or power, but a desire to deny Germany the ultimate weapon. Szilard, a Hungarian Jew and an early refugee from Nazism, had long been convinced that Hitler would bring Europe to war. He offered his patents to the British Army, which turned him down with the curt observation that 'there appears to be no reason to keep the specification secret as far as the War Office is concerned'. The Navy accepted Szilard's patents, but never exploited them.

Releasing the enormous energy of the atom – either as a devastating explosion or as a controlled source of industrial power – was not a new idea. Szilard had first encountered it in H. G. Wells's book, *The World Set Free*, but as far as the leading scientists of the day were concerned, it was physically impossible. Rutherford did not deny that it was possible to change the structure of atoms. In fact, Rutherford himself had been the first to do so when in 1919 he succeeded in transforming a few nitrogen atoms (which have seven protons in their nuclei) to atoms of oxygen (which have eight protons) by bombarding the nitrogen atoms with alpha particles.

Szilard, however, envisioned nuclear fission – a much more radical form of atomic alteration. In fission, an atom does not simply add or lose a proton or two. It is split into two or more fragments, and some of the immense 'binding energy' that holds the positively charged nucleus together is released. The larger the atom, the more energy is required to keep the many protons in its nucleus from flying apart, and thus the more energy that would be released if its nucleus could split. It was, said Rutherford, an idea that could never be tested. Speeding alpha particles might be able to knock a proton out of a small nucleus, but the positively charged alpha particles were repulsed by the strong positive charge of atoms with large nuclei. The larger the atom, the more it would repel any alpha particle that approached it. Thus, fission seemed only a dream of science fiction writers.

Then, in 1932, James Chadwick, working under Rutherford at the Cavendish Laboratory in England, discovered the neutron, which is essentially a proton plus an electron. As heavy as a proton, but without an electrical charge, the neutron was the perfect bombardment tool, able to fly into the nucleus of an atom and break it apart. But most scientists were still unable to conceive of a simple neutron overcoming the enormous force that binds an atom together. Though Enrico Fermi split uranium atoms by bombarding them with neutrons in 1934, four years passed before he could bring himself to believe that he had done so. Fermi, and other physicists who conducted similar experiments, searched high and low for other explanations for the results they could not accept. In 1934, only one person, a young German physicist named Ida Noddack, saw what was really happening. 'When heavy nuclei are bombarded with neutrons,' she wrote in reference to Fermi's work, 'the nuclei in question might break into a number of larger pieces which would no doubt be isotopes of known elements.' Otto Hahn, who was a friend of Noddack's, gallantly refused to publicize her assertion because, he told her, 'he did not want to make me look ridiculous, as my assumption of the bursting of the uranium nucleus into larger fragments was really absurd'. Not until 1938, when Hahn and Fritz Strassmann reluctantly reached what Hahn called the 'horrifying conclusion' that they too were splitting uranium atoms in their Berlin laboratory, did the scientific community recognize that nuclear fission was possible.

The fact that the discovery of fission had been made in Germany deepened Szilard's fear that Hitler might one day possess a weapon that would give him control of the world. In 1939, Szilard, who by then was working with Fermi at Columbia University, persuaded most of his colleagues to agree to a voluntary self-censorship on information about fission in order to deny crucial information to the Germans. In early March 1939, Fermi, Szilard, and colleagues at Columbia proved theoretically that a chain reaction was possible, but did not publish their finding because of its military applications. At about the same time, Irène and Frédéric Joliot-Curie, the daughter and son-in-law of Madame Curie, and their co-workers in Paris made a similar discovery. The Joliot-Curies rushed their results into print. The French team's announcement that a chain reaction was feasible appeared on 15 March 1939, the very day that Germany invaded Czechoslovakia. With the demise of the censorship agreement, more than 100 scientific papers on nuclear fission were published in 1939 alone, along with many sensational newspaper stories about the practical applications of atomic energy. In 1940 *Collier's* magazine described the 'atomic future' in euphoric terms. According to the author, Rudolph Langer, a physicist at the California Institute of Technology, fission would enable agriculture and cities to move underground, so that the surface of the earth would be devoted to parks and greenery. Fission-powered automobiles and airplanes would soon carry people around the world in a few hours. 'We are about to enter a period of unparalleled richness and opportunities for all,' Langer wrote. 'Privilege and class distinctions and the other sources of social uneasiness and bitterness will become relics because things that make up the good life will be so abundant and inexpensive. War itself will become obsolete.'

Szilard had a far less rosy vision of the future. He feared the Germans would get a monopoly on the atomic bomb. In August 1939, at Szilard's urging, Albert Einstein signed the famous letter to President Roosevelt warning of the destructive potential of nuclear fission, which led two months later to Roosevelt's appointing an Advisory Committee on Uranium to look into the feasibility of an atomic bomb. But Szilard's sense of urgency was not widely shared: the committee's work proceeded slowly, as did that of similar groups in Britain and Germany. Then, in March 1940, a three-page memo written by Otto Frisch and Rudolf Peierls, two German refugee scientists living in England, put new life into the British bomb effort. Frisch and Peierls argued that as little as five kilograms of uranium would be needed to construct a bomb with the destructive force of several thousand tons of dynamite. They also warned, for the first time, that the radioactivity produced by such a bomb would be dangerous to human life.

The reinvigorated British bomb group, involving scientists from Birmingham, Cambridge, Liverpool and Oxford, called itself the MAUD Committee. The name was taken from an incident involving the Danish genius, Niels Bohr. After the Germans invaded Denmark in April 1940, the physicist Lise Metiner, Otto Frisch's aunt, sent a cable from Sweden with assurances that Bohr, whom she had recently seen, was safe. The cable ended with the words, 'PLEASE INFORM COCKCROFT AND MAUD RAY KENT.' John Cockcroft was an eminent British scientist, but what of 'Maud Ray Kent'? Frisch ingeniously concluded that the words were an anagram of 'radyum taken', a message that the Nazis had confiscated the Danish supply of radium. Actually, as Frisch later learned, Maud Ray had been governess to the Bohr family before she retired to Kent. When an unrevealing title was needed for the British bomb project, Frisch self-mockingly proposed MAUD. At the end of July 1941 the MAUD committee produced a report stating that it was possible to make a uranium bomb, and that the first one could be ready by the end of 1943.

In Britain the MAUD report disappeared into the overloaded wartime bureaucracy but, across the Atlantic, the report galvanized the Americans into action. In October 1941, before the United States had even entered the war, Roosevelt ordered the Office of Scientific Research and Development to develop an atomic bomb. The key job of achieving the first chain reaction went to the Nobel Prize-winning physicist, Arthur Compton, at the University of Chicago. Compton wasted no time in bringing in scientists, and even whole research projects, from other universities. Fermi came from Columbia to design the first nuclear reactor, the structure in which, it was hoped, chain reactions – still only a theory – would take place. With the flair for unrevealing names that was to characterize the entire bomb-building effort, Compton designated his project the Metallurgical Laboratory, the Met Lab for short. In January 1942, work on the bomb began.

Within 18 months the bomb project had mushroomed from a

university research effort involving 45 people, into a huge industrial operation run by the Army. The Met Lab was only one part of the whole bomb project, called the Manhattan Engineer District after the army field office under whose aegis it originally came. The Manhattan Engineer District, renamed the Manhattan Project after the war, was the biggest engineering task ever undertaken. By the end of the war, it comprised three specially built cities with a total population of almost 150,000; 39 plants and research facilities throughout North America; more than 40,000 workers; a budget of $2.2 billion – and an obsession with secrecy. The Project also had its own code: plutonium was 'product'; the atomic bomb was 'the gadget'; and radiation was 'the special hazard'.

Until the 1940s, ionizing radiation was a specialized tool, used almost exclusively by a few hundred doctors and researchers. The Manhattan Project drastically changed that. During the war radiation was produced and used on an industrial scale. The risks of using it were experienced not merely by a small number of highly trained technicians, but by tens of thousands of workers with no understanding of those risks. It was up to the experts of the Manhattan Project's Health Division to protect these workers.

The Health Division was established when the Met Lab was six months old. The scientists working on the project were not particularly concerned about the large-scale use of familiar radiation-producing materials and equipment, but they were apprehensive about the completely new and unknown radioactive materials that the project would produce and with which they would have to work. General Leslie Groves, who took charge of the Manhattan Project for the Army in September, 1942, had a more positive view of the effects of radiation. 'They say it is a very pleasant way to die,' he told a postwar Senate committee on atomic energy. Nonetheless, according to Compton, 'our physicists became worried. They knew what happened to the early experimenters with radioactive materials. Not many of them had lived very long. They were themselves to work with materials millions of times more active than those of these earlier experimenters. What was their own life expectancy?' It was to allay such fears that Compton established the Health Division.

Robert Stone, chairman of the Department of Radiology of the University of California Medical School in San Francisco, was chosen to lead the Health Division, which was composed of three sections. The health physics section had the job of detecting radiation and designing safeguards. The medical section examined workers. The biological research section studied the health effects of the materials to which workers were exposed. A small cadre of scientists – Stone had only eight radiation protection specialists, for example – was assisted by military recruits who received on-the-job training. Stone's men had two very different tasks: they had to protect a large population against the well known hazards of gamma radiation, and they had to devise protective measures against sources of radiation so new that their hazards were still unknown.

In 1934, the US Committee on X-Ray and Radium Protection decreed that workers could tolerate one-tenth of a roentgen of X-rays or gamma rays a day. Six years later, however, several members of the US committee proposed that the standard be lowered by a factor of five in response to the accumulating evidence that any amount of radiation, no matter how small, can cause genetic damage, injuring future generations. Though broadly supported, the suggestion drew some criticism in the committee. Gioacchino Failla objected on the grounds that if genetic damage were to be a consideration for standard-setters, then logically no radiation exposure should be allowed. 'If we bring in genetic criteria then there is no limit at all and 0.02 [roentgen per day] is just as arbitrary as 0.1,' he wrote to his fellow committee members. The proposal, said Lauriston Taylor, 'was lost in another crack' when war broke out.

Stone too had his doubts about the validity of the 1934 standards. 'It was evident,' he said, 'that these rested on rather poor experimental evidence.' Whatever his reservations, with the Manhattan Project already under way, Stone had no choice but to adopt the existing standards, but he hoped to develop more reliable standards as the project continued. The Health Division sponsored research at the National Cancer Institute in Washington, in which large numbers of mice were subjected to low-level gamma radiation for long periods of time. The study's conclusion further undermined confidence in the existing standard for external radiation. In Stone's words, the studies showed that 'the one-tenth of a roentgen [limit] has no greater factor of safety than ten, and that further studies might reveal that even this factor is too large'. Rather than setting a new, lower limit, however, Stone simply renamed it. In the new terminology, the 'tolerance dose' became the 'maximum permissible exposure' – a wording, Stone argued, that emphasized the need to keep exposures as far as possible below the ceiling.

Mild though it was, this change met with resistance from the project's military overseers. Stafford Warren, medical adviser to General Groves, clashed with Stone on the issue, arguing that it was wasteful to lower exposures beyond the established limit without evidence of 'definite biological changes which are immediately evident or are of prognostic importance to health'. Warren's deputy, Hymer Friedell, warned that subjecting workers to intense monitoring might actually increase danger by engendering 'untoward psychological effects in individuals working with a new and unknown material'. Alarmed or inquisitive workers could, he pointed out, jeopardize the deep secrecy enshrouding the project.

Stone's immediate concern, however, was simply meeting the existing limit, not trying to beat it. It was difficult to monitor workers or work areas, because monitoring equipment was scarce and rather primitive. The one readily available monitoring tool was the pocket ionization chamber, a fountain-pen-sized gas-filled tube that responded to radiation. All workers in high radiation areas carried the tubes in their pockets. Because the ionization chambers tended to exaggerate exposure, each worker carried two, and whichever gave the higher reading was ignored. Even so, the pocket monitors were hard to interpret – especially for the hurriedly trained enlisted men who made up the bulk of the monitoring force. In time the health physics section developed specially filtered photographic film that gave fairly reliable dose readings. The film was attached to the security badge worn by all project workers – a forerunner of the film badge that is now standard issue to workers throughout the nuclear industry. Workers who exceeded the limit were interviewed, along with their supervisors, and ordered to reduce their exposure.

Lack of money and staff, however, restricted the monitoring programme to those most at risk: workers whose jobs were likely to involve only moderate exposure, or no exposure at all, were not closely watched. 'We do not have important legal evidence in case of future claims against this project,' admitted Louis Hempelmann, head of the Health Group at Los Alamos, New Mexico, where the bombs were to be designed and produced. Officials were aware that even those who had been closely monitored and received a clean bill of health might develop problems later. John Wirth, medical director of the project's Oak Ridge laboratory in Tennessee, described with satisfaction the

Project's radiation monitoring programme as 'more search than discovery', but he also warned of the possibility of 'the unexpected appearance of dangerous changes months or years after exposure'.

The Manhattan Project provided the first chance to study radiation effects on a large population. 'The whole clinical study of the personnel is one vast experiment,' noted Stone. 'Never before has so large a collection of individuals been exposed to so much irradiation.' The military wanted the Health Division's monitoring and research work to help strengthen, in Stafford Warren's words, 'the government interests' in case of lawsuits or financial demands by workers claiming to have been injured by their work on the bomb. This attitude angered Stone. He argued that studies should concentrate upon the effects of long-term exposure to the new forms of radiation that the project was introducing into the world.

Warren's views ultimately prevailed: most project research addressed immediate wartime problems. Nevertheless, the Health Division and its associated researchers discovered a great deal about how to monitor and measure radiation, and about its immediate and delayed effects on humans and animals. In addition to the National Cancer Institute's mice studies, the Health Division sponsored a series of human experiments in New York, Chicago, and San Francisco, in which people suffering from incurable diseases were subjected to large doses of radiation, so that doctors could study the delayed effects of irradiation for the few remaining months or years of the subjects' lives.

The Health Division's most challenging job was protecting workers against entirely new radioactive materials with which neither the division nor anyone else had any experience. The greatest hazard was plutonium, a previously unknown radioactive metallic element. In 1940 Glenn Seaborg and his colleagues produced a tiny quantity of plutonium in the University of California's Berkeley cyclotron by bombarding uranium with neutrons. Subsequently, scientists found minute amounts of plutonium in nature, but virtually all the plutonium now in existence is man-made.

The Manhattan Project also created hundreds of new and unknown fission products – radioactive isotopes created when plutonium, uranium, and other elements undergo fission. In the immediate aftermath of a fission reaction, many of the newly created isotopes are so radioactive that they decay within seconds or minutes into other isotopes, which themselves decay into others, and so on. There may be hundreds of fission products in these decay series. The shortest-lived ones are the most radioactive, but the most dangerous are the 20 or so that survive long enough to contaminate humans or the environment. These long-lived fission products include strontium–90, caesium–137, and krypton–85. Another class of radioactive substances that were created for the first time during the Manhattan Project are the 'neutron-activation products'. Activation products are materials inside and near a nuclear reactor that become radioactive by 'capturing' neutrons released during the nuclear reaction. Cobalt–60 and carbon–14 are two of the hundreds of radioactive substances created this way.

The biggest mystery was how these man-made radioactive products would affect the body. External radiations (gamma and X-rays) were fairly well understood, and there were accepted methods of protection against them. But many of the new materials were certain to emit alpha and beta particles which, though unable to penetrate deep into the body from outside, can cause serious damage if taken into the body. The type and amount of damage caused by radioactive materials inside the body depends not only on the nature and strength of their radioactivity, but also upon their chemical properties: the body tissues, if any, in which they tend to concentrate; whether they are soluble and may enter the blood stream; whether they are influenced by the presence of other chemical elements, and so on. It was too soon to have the answers to these questions for the radioactive substances being created by the Manhattan Project. Nonetheless, the Health Division had somehow to ensure that these new hazards were kept under control.

After the war, Robert Stone described the obstacles of ignorance his staff faced:

> The exact nature and extent of the hazards to be encountered were not known. In general it was realized that the pile [the nuclear reactor] in which the chain reactions would take place would be a far greater source of fast and slow neutrons and gamma rays than could have been dreamed possible in earlier periods. It was estimated that the pieces of uranium that would have to be removed from the pile after fission had occurred would contain materials far more radioactive than any that had been encountered in the radium industry. It appeared certain that not only the uranium but also any other material that would be removed from

such a pile would be extremely active in producing alpha, beta, or gamma rays. The chemical process of separating the plutonium from the other extremely radioactive elements was recognized as another tremendously hazardous procedure. The effect that plutonium itself might have on the workers was unknown.

Scientists did know that plutonium – like radium – emits alpha particles. Thus, the standard set in 1941 for radium provided a guideline for plutonium. That standard recommended that anyone found to have accumulated more than one-tenth of one microcurie of radium in his body should 'change his occupation immediately'. Based on its alpha activity, plutonium was initially estimated to be only about one-fiftieth as hazardous as radium, so the plutonium limit, or maximum permissible concentration, was set at five microcuries, 50 times the limit for radium. Enforcing this standard was almost impossible, however, because there was no practical way to measure the amount of plutonium in a person's body. Radium can be detected by its gamma emissions, but plutonium can only be detected by its alpha activity, and there were very few alpha monitors in existence in 1944. Even if it were possible to figure out how much plutonium a person had inhaled or swallowed, no one knew how much of that remained in the body and how much passed out immediately. There was therefore no way to determine a worker's 'plutonium burden'. This uncertainty caused Stone to insist that 'the only amount of product [plutonium] that should be inhaled is none at all'.

The project's over-stretched instrument group, headed by the British physicist, Herbert Parker, was unfortunately unable to develop a good portable alpha counter before the end of the war. Monitors were therefore forced to use what they called the swipe method to obtain rough estimates of alpha contamination on work surfaces. This involved wiping a work surface with a one-inch-square piece of filter paper and then putting the paper into an alpha-counting machine. To get an idea of how much plutonium dust was in the air, monitors took 'nose counts', a variation of the swipe method in which the filter paper was used to wipe a worker's nose and then submitted to the alpha counter.

Parker also had the job of developing ways of measuring the new substances. One problem was that the existing radiation unit, the roentgen, measured only gamma rays and X-rays, whereas the workers that Parker and his colleagues were trying to protect were also exposed to alpha, beta, and neutron radiation. What was needed was a unit that could be applied to all forms of ionizing radiation. Parker came up with the idea of describing a dose of radiation in terms of the amount of energy it deposits in one gram of human tissue. Parker called the new unit a rep; later it was slightly redefined and renamed the rad. The biological impact of a rad, however, varies according to the type of radiation involved. One rad of alpha radiation is believed to have a biological impact from ten to twenty times greater than one rad of gamma radiation. This is due to the fact that alpha radiation transfers its energy to a target so much more rapidly than other radiations that its biological impact is more concentrated. To deal with this, Parker invented yet another unit, the rem, which takes into account the biological effectiveness of different types of radiation. Thus, a dose of twenty rems is equal to twenty rads of gamma radiation, but only one or two rads of alpha radiation.

Not until 1944, when plutonium began to be produced in quantity, could researchers begin to study its health effects. The first batch of plutonium available for research went to Joseph Hamilton and his colleagues at the University of California at Berkeley, who were under contract to the Health Division. Their findings on plutonium validated the chemists' fears. Like radium, plutonium is a bone-seeker, but rather than spreading throughout the bone, plutonium tends to settle on the surface, near the blood-forming bone marrow. Plutonium also remains in the body longer than radium. Hamilton informed Manhattan Project officials that, because of its behaviour in the body, plutonium is actually five or ten times *more* toxic than radium. Eventually, in 1945, project officials lowered the maximum permissible concentration of plutonium in the body from five micrograms to one microgram – still ten times higher than the radium limit.

The delicate task of purifying the raw plutonium so that it could be used in a chain reaction fell to the chemists at Los Alamos. Not one of them had ever seen plutonium before, but they knew that it was likely to be as dangerous as radium, if not more so, and they knew what had befallen the radium dial painters. An indication of how seriously the hazard of the unknown substance was taken was the adoption of a policy of 'immediate high amputation' if plutonium entered a cut or scratch on a worker's body. The chemists refused to work with plutonium without extra insurance. The Manhattan Project had worker insurance, but Cyril Stanley Smith, the chief metallurgical chemist at Los Alamos, called it 'inhumane, unethical, and unfair' because it covered only illnesses or disabilities that appeared within 90 days of an accident or 30 days of leaving the project, even though radiation damage might take thirty years to surface. In the end, the Army and the University of California set up a secret million-dollar fund for the plutonium chemists.

The most risky and untested undertakings at Los Alamos, for which in many cases even the equipment was highly experimental, were banished to canyons far away from the main work area. In case of an accident there would be fewer casualties in these isolated spots, but they were almost impossible to monitor. According to Louis Hempelmann, chief of the Los Alamos Health Group, there were a number of serious accidents with equipment that malfunctioned and exposed scientists to 'considerably more radiation dosage than was desirable'. In one month, August 1944, there were two major plutonium spills at Los Alamos. Also in that month, ten milligrams of plutonium exploded in the face of a Los Alamos chemist, who consequently swallowed an unknown amount. Health officers had no way of telling how much plutonium remained in his body or how much damage it had done. Shocked by their own ignorance, they began looking for ways of measuring plutonium in the body. They soon devised a way of detecting one-trillionth of a gram of plutonium in a litre of urine, and in early 1945 began testing workers regularly. The tests showed that some workers already had too much plutonium in their bodies – a condition for which there is no remedy. Plutonium–239 has a half-life of more than 24,000 years. Once plutonium has settled in the body, it remains there, continuing to irradiate the surrounding tissues long after the person has died.

Stone freely admitted that it was 'impossible to keep all sources of radiations under control'. The pressure of work was intense, much of the lab equipment was improvised, the hazards were invisible, and some material was so radioactive that a single drop could contaminate an entire building. John Wirth at Oak Ridge wrote that 'minute invisible fragments might make an entire building uninhabitable until it had been decontaminated . . . It is always amazing what widespread contamination can be caused by a minute quantity of hot material once it has been allowed to get outside a container. No less amazing is the ease with which it seems to get about, as though it were a living creature trying to spread itself anywhere.'

Plutonium was so valuable and so scarce that losing even a tiny speck of it was an expensive technical setback, as well as a health hazard. After the war, an equipment inspector at Oak Ridge's Y–12 plant, where uranium—235 for the first bomb was produced, told Daniel Lang of the *New Yorker* about strange activities he had witnessed while working there during the Manhattan Project: 'the employees got down on their knees every now and then to look for tiny bits of metal. In their hands they held Geiger counters . . . Sometimes the counters would lead them to a tiny orange or black speck on someone's white uniform. The speck was very valuable. The uniforms of certain workers were always chemically treated before they were laundered or junked to make sure that none of those precious bits would go down the drain.'

Ted Lombard, assigned as a GI to Los Alamos during the Manhattan Project, also remembered less-than-ideal working conditions:

> I'll tell you what the working conditions were at Los Alamos . . . We used to go to Fort Douglas, Utah in ambulances, pick up uranium and plutonium. We carried dosimeter badges in our pockets because you couldn't display them, that was not permitted. When we arrived back at the technical area, the lieutenant would pick up the dosimeters and disappear with them. Then we would proceed to unload the uranium and plutonium barehanded . . . The fumes and dust were constantly in the air; there was no ventilation system. The dust was on the floor. Uranium chips would be in your shoes that you continued to wear. You went to eat with the same clothes on, you went into the barracks with the same clothes and sat on the beds . . . Contamination was rampant and there was little or no protection, particularly for the GIs.

Lombard now suffers from fibrosis of the lungs and an undiagnosed skin disease, and severe damage to the bone marrow, the blood-forming organ. Four of his five children, born after he worked at Los Alamos, have severe medical problems, including neuromuscular and blood disorders. His eldest son, the only one of the five who was born before Lombard entered the service, is healthy. Lombard attributes his and his children's health problems to his radiation exposure at Los Alamos, but the Veterans Administration rejected his repeated claims for compensation, saying that the illnesses were contracted after he left the service. Officials say that Lombard's medical records are missing and that there is no record of his radiation exposure while at Los Alamos. Recently, the VA relented, agreeing to pay him some compensation, while specifying that his injuries are not radiation-related. Regarding his children's ailments, the VA noted that 'there is no provision in laws and regulations for compensation based upon birth defects in children claimed to be caused by the genetic injury to the parent-veteran in service'.

Despite the occasional accidents, health officials were pleased with their handling of the 'special hazard'. 'The Project was fraught with tremendous potentialities for over-exposure to radiation, but in actuality there were no over-exposures of serious import,' Wirth reported after the war. In 1951 Stone could say with satisfaction, 'no one lost his life because of any of the hazards peculiar to the project'. In fact, officials felt that radiation, the 'special hazard', was not the most serious or widespread health problem. Toxic chemicals may have posed a greater risk to the workforce than radiation. 'It actually took a larger medical organization to handle the general industrial work than it took to cope with the special hazard problem,' said Wirth. Most of the project's medical staff received only a 'brief indoctrination . . . in the recognition of radiation sickness', according to Warren, and, luckily, they never had to use it.

CHAPTER SEVEN

Getting to Know the Bomb

In the final weeks of the war, the Manhattan Project was suddenly presented with an unanticipated radiation hazard – fallout. Though the German refugee scientists, Otto Frisch and Rudolf Peierls, had suggested in 1940 that an atomic bomb might release dangerous amounts of radioactivity into the environment, the Manhattan Project scientists had been more concerned with building an atomic bomb than with its possible radiological effects. As Hymer Friedell, the project's deputy medical officer said, 'The idea was to explode the damned thing . . . We weren't terribly concerned with the radiation.'

In June of 1945, however – just one month before the testing of the first atomic bomb, code-named Trinity – new calculations suggested that fallout from a successful explosion could force the evacuation of hundreds of civilians from around the remote test site at Alamogordo, New Mexico. This was an unwelcome extra burden in the hectic days before the test. To the extent that the requirements of radiation safety dovetailed with the project's overriding need for secrecy, safety was served. But when the two clashed, secrecy prevailed. For example, Groves rejected the idea of preparing the nearby ranchers and towns-people for evacuation, a stance that considerably complicated the Army's emergency planning for the test.

Trinity's radiological safety would depend to a large extent upon the weather during the test shot. If it rained during or just after the shot, the radioactive materials produced by the explosion would be washed down out of the sky, instead of being carried high into the upper atmosphere, or being widely dispersed before returning to earth. How much fallout there would be, how thoroughly it would be dispersed, and where it would land, would depend largely on wind conditions during and after the shot. For technical reasons, the bomb would not be ready before 15 or 16 July. Weather forecasts pinpointed 18 or 19 July as the earliest good days for the shot, but President Truman's aides told Groves that the bomb had to be tested by 16 July, the first day of the

Potsdam Conference – whatever the weather. That decision increased the likelihood of Trinity's contaminating the test site and a large area beyond.

Protection measures within the test site centred on banishing most observers to a site 20 miles to the northwest of ground zero. Scientists and others whose jobs required them to be closer, sheltered in specially built bunkers about five and a half miles from ground zero. Many scientists wanted to go into the danger zone shortly after the test to retrieve their data; they were required to sign releases agreeing that it was 'the individual's responsibility' to limit his own radiation exposure. The health group strongly suggested that people restrict themselves to a total exposure of 5 roentgens.

The safety of everyone outside the test site depended upon Stafford Warren, the Manhattan Project's chief medical officer, who would watch the test at Alamogordo, and his deputy, Friedell, who jokingly remarked that his post, a hotel room at Albuquerque, 100 miles away from the test site, would ensure that any accident 'would only happen to half of us'. In the aftermath of the test, Warren and Friedell would have the power to order the area around Alamogordo evacuated. How much radiation would warrant such a drastic action? Warren felt that a total exposure of 60 roentgens (r) of gamma radiation over a two-week period would be safe, and that 'even 100 r would not be harmful provided there would be no further exposure'. The health section finally set the triggers for evacuation at 15 r per hour or 75 r in two weeks. The limit for project workers at the time was 1 r in two weeks.

In the weeks preceding the test, Army Intelligence agents scoured the area around Alamogordo, trying to record the location of every person living within 40 miles of ground zero. The area was remote, but not empty. There was a ranch only 15 miles from ground zero and, five miles further, a town. To gauge the strength of the fallout cloud after it left the test site, a fleet of radio-equipped cars fitted with alpha and gamma monitors would patrol a 150-mile radius from ground zero. The test site itself was ringed with alpha air samplers, nicknamed Sneezies, that health group technicians had adapted from Filter Queen vacuum cleaners. A 144-man unit, the Army Evacuation Detachment, was formed and provided with enough food and equipment to set up a temporary camp for 450 people. Evacuees would be told that an ammunition dump containing poisonous gas shells had blown up. In case things really went wrong, Groves was authorized to declare martial law 'over as large an area as might be necessary'. The Army took seriously the possibility of lawsuits stemming from fallout injury. Its agents sent film badges by registered mail to all local post offices. Both the badges and the car readings would provide valuable legal evidence in case of any future claims.

Trinity day, Monday 16 July, dawned overcast, with a thunderstorm threatening. The shot was postponed for one and a half hours. Finally at 5.10 it was announced that 5.30 a.m. would be zero hour. At 29 minutes and 15 seconds past five, Trinity was detonated. It exploded spectacularly, sending radioactive dust and debris into the heavens. The planners' worst nightmare – having to crawl out and dismantle a dud atomic bomb – was not realized. The news was flashed to Washington, and thence to Potsdam, where Truman made his veiled threat to the Japanese.

Immediately after the explosion, scientists began taking radiation readings at the forward bunker positions, which were more than five miles from ground zero. Readings were very low. Most of the so-called prompt radiation – the radiation that reaches the ground at the time of the explosion, as opposed to the radiation carried aloft during the explosion – decayed in less than one second. The bomb crater itself, however, was much hotter than expected. Tanks equipped with earth-sampling instruments approached the crater one and a half hours after the test, and found it too radioactive for the project's instruments to measure. Estimated readings were 600 to 700 r per hour. Planners had not allowed for the fact that neutrons released in the explosion would react with sodium and other elements in the soil and make them radioactive.

After the blast, the fallout cloud travelled northeast. Two health group members, physicists Joseph Hirschfelder and John Magee, were sent after it along Highway 380 which ran parallel to the test site's northern edge. At an isolated crossroads called Bingham they found a general store tended by an old man. As Hirschfelder later remembered it: 'He looked quizzically at us (John and I were wearing white coveralls with gas masks hanging from our necks). Then he laughed and said, "You boys must have been up to something this morning. The sun came up in the west and went on down again."' Hirschfelder and Magee detected fallout at Bingham, but concluded that there was not enough to be dangerous. The two physicists moved on to a nearby Army outpost, where a group of soldiers were about to begin a post-bomb breakfast of T-bone steaks. 'When we arrived they were just roasting the meat and it smelled delicious. However, at the same time the fallout arrived – small flaky dust particles gently settling on the ground. The radiation level was quite high.' The reading was 2 r per hour – 20 times the then-current daily limit for radiation workers. The soldiers were ordered to bury the steaks and leave the area.

Varying wind-patterns and terrains meant that the fallout did not settle evenly. At 9.00 a.m., Magee got a reading of 15 r per hour at a spot northeast of Bingham that was about 30 miles from ground zero. The highest readings were found near by in a ravine that was soon nicknamed Hot Canyon. Luckily, the Army maps showed that no one lived in the area. There was also a brief period of concern about the town of Carrizozo, near the test site's eastern border, but radioactivity levels there dropped quickly and no evacuation orders were given. By Monday evening, everyone relaxed: the crisis had passed.

On Tuesday, Friedell and Hempelmann, still puzzled by the high readings at Hot Canyon, drove out there. To their horror, they discovered an adobe house only one mile from the highest readings. An elderly couple named Raitliff and their ten-year-old grandson lived in the house. Hidden from the road, the building had been overlooked by the Army's census-takers. Mr Raitliff subsequently told Hempelmann that the ground had been 'covered with light snow' on the day of the shot, and that at dawn and dusk for several days after 'the ground and fence posts had the appearance . . . of being "frosted"'. Later a second family, the Wilsons, was also discovered near by. The families' live-stock and pet dogs were burned and bleeding and had lost their hair, but the families, who appeared healthy, were not evacuated. Instead, Hempelmann and other project health officials visited them periodi-cally for about six months. In December, monitors found readings of external gamma radiation at the Raitliff ranch were 19 milliroentgens per hour, or almost half a roentgen per day – five times the standard for radiation workers. Allowing for the shielding provided by the thick adobe walls of their homes, Hempelmann estimated that the Raitliffs received no more than 47 r of gamma radiation in the two weeks after Trinity, and that the Wilsons received about three-quarters of that. External exposure from beta particles and internal exposure from inhaled alpha and beta particles was not measured, though many fallout products are alpha and beta emitters. Hempelmann's last visit to the Raitliffs was in November. They had been by then joined by their two-year-old niece, and Hempelmann felt that they were all in good health.

The chief sufferers from Trinity, officials concluded, were animals belonging to Raitliff, Wilson, and other ranchers, that grazed in the high fallout area. The Army estimated that 600 cattle had been injured by the Trinity fallout. Robert Stone attributed the cattle injuries, mostly burns and loss of hair, to external beta radiation. He estimated that the animals had received 'probably about 20,000 roentgens', possibly as many as 50,000. Humans caught in the fallout would have received a much smaller dose, by virtue of their tendency to wear clothes and take baths, but the Army feared that this distinction would be lost on the public.

Worried about lawsuits, Robert Oppenheimer, the brilliant physicist who oversaw the making of the bomb, ordered the Health Group's Trinity reports to be held in the strictest secrecy. They were kept separate from other Trinity reports and could be released only with Oppenheimer's personal approval. After the war, when the Manhattan Project and the Trinity test became public knowledge, a local rancher did file a damage claim against the Army. To help hush up the incident, the Army eventually bought 75 damaged cattle and sent them to Oak Ridge, where their descendants are still studied.

There was no real effort made at Trinity to gather information on the behaviour of fallout, or its effects on humans and animals. Resources were concentrated on simply getting the bomb to work. Only one biological experiment was sanctioned by Groves, and it was a macabre failure. Some mice were tied by their tails to a piece of rope and hung near the explosion, but they died of thirst in the desert heat before the bomb was detonated. Nor was there any organized tracking of the fallout cloud after the first day, when the cloud passed into Colorado and was assumed to have dispersed safely. Three years afterwards, a team of monitors belatedly tried to trace the path of the Trinity fallout. Their main clue came from disgruntled Eastman Kodak customers who complained that they were sold fogged film. An investigation eventually discovered that the film had been packed in paper made of straw that had been washed in water from the Wabash River which had been contaminated with fallout from the explosion at Alamogordo, more than 1,000 miles away.

Overall, health officials were satisfied with Trinity. There had been some close calls, but though animals had been injured, no humans on or off the test site had been harmed. Despite the widespread fallout, the project's secrecy had been maintained. The need for secrecy ended on 6 August 1945, when the world learned that a place called Hiroshima had been destroyed by what Truman called 'a harnessing of the basic power of the universe'.

Two days after the bombing of Hiroshima, many American newspapers carried a story warning that radiation from the atomic bomb might make Hiroshima uninhabitable for 70 years. The story was based on an interview with a former Manhattan Project scientist, Dr Harold Jacobson of Columbia University. Jacobson predicted that the long-term radiation effects of the atomic bomb would be even worse than the immediate damage from blast and fire. The story warned that 'Hiroshima will be a devastated area ... for nearly three-quarters of a century ... Rain falling on the area will pick up the lethal rays and will carry them down to the rivers and the sea. And animal life in these waters will die ... Investigators in a contaminated area will become infected with secondary radiation which breaks up the red corpuscles in the blood. People will die much the same way leukemia victims do.'

Jacobson's dire predictions elicited an immediate reaction from his former employers. Manhattan Project officials denigrated him to reporters as a lowly technician. Robert Oppenheimer publicly ridiculed his comments. The War Department pressured him to retract them. After intensive questioning by the FBI, Jacobson announced that his statements were not based on inside information, but were simply his personal opinion.

Meanwhile Tokyo Radio began announcing that people who entered the city after the explosion were dying of mysterious causes. American officials dismissed the allegations as propaganda intended to imply that the US had used an inhumane weapon. Determined to put the rumours and accusations to rest, General Groves ordered a team of Manhattan Project doctors and technicians to the two bombed cities, Hiroshima and Nagasaki. According to Groves's deputy, General Thomas Farrell, 'our mission was to prove that there was no radioactivity from the bomb'. The delegation arrived in Japan a month after the bombings. After a four-day reconnaissance, Farrell informed Groves that there was no trace of radioactivity in either city. Most of the survey team stayed longer in Japan, under the command of Stafford Warren. They did find some radioactivity, but, according to Warren, it was 'below the hazardous limit; when the readings were extrapolated back to zero hour, the levels were not considered to be of great significance'. Warren concluded that lingering radiation was not a problem for the occupying troops or Japanese citizens.

On 5 September, shortly before Farrell's team arrived at Hiroshima, London's *Daily Express* newspaper carried a front-page story headlined 'I WRITE THIS AS A WARNING TO THE WORLD' by Wilfred Burchett, the first newsman to enter Hiroshima without an Army escort. 'In Hiroshima, 30 days after the first atomic bomb,' he wrote, 'people are still dying, mysteriously and horribly – people who were uninjured in the cataclysm from an unknown something which I can only describe as the atomic plague.' American officials in Tokyo and Washington strenuously denied Burchett's charges.

A group of American journalists who entered Hiroshima under US army escort around the same time as Burchett reached a different conclusion. The *New York Times* headlined its account, 'No Radioactivity in Hiroshima Ruin.' The general conclusion was that tales of radiation poisoning in the bombed city were groundless. Concern about radiation played no part in the public feeling about the atomic bomb. In the weeks after Hiroshima, radiation was not mentioned in a single one of more than 200 letters about the atomic bomb published in American newspapers.

In this important respect, the world underestimated the power of atomic bombs. Almost everyone thought that atom bombs, for all their awe-inspiring destructive force, were simply super-powerful versions of conventional bombs – 'just another piece of artillery', as Truman put it. Atomic bombs do kill by blast and fire, like ordinary bombs. But unlike ordinary bombs, unlike any weapon the world had seen before, atomic bombs also kill by silent, invisible radiations that persist long after the blast and fire have ended.

CHAPTER EIGHT

Postwar Standards

'The atom had us bewitched,' one observer remarked of the early postwar days. 'It was so gigantic, so terrible, so beyond the power of imagination to embrace that it seemed to be the ultimate fact. It would either destroy us or bring around the millennium.' There was indeed a widespread feeling that the atomic bomb had set mankind at a cross-roads. As the editors of the *Woman's Home Companion* put it, 'The choice is as simple as this – if we have peace we can have paradise; if we have war, we could face doomsday.' Many commentators used biblical comparisons: mankind's possession of the secrets of the atom gave it the opportunity to redeem itself, to make up for the sin that had cast it out of the Garden of Eden. If the new power was used for good, the human race would enter a new Eden; if it was used for war, all life on earth would be destroyed in a nuclear holocaust.

The *New York Times* urged its readers to look on the positive side of atomic power: 'Beyond the veil of dust and smoke and searing death which was Hiroshima, there are many fascinating fields of speculation for the use of atomic power for beneficent rather than destructive ends.' Those were welcome words to many people – most of all, perhaps, to the scientists who had helped create the bomb and who wanted desperately to believe that the great power they had unleashed could be a force for good as well as evil.

The American Psychological Association warned in June 1946 that 'the possible benefits of atomic energy must be emphasized and developed' so that 'the atmosphere of demoralizing fear which surrounds the phrase "atomic energy" can be reduced'. Otherwise atomic-phobia could destroy the nation. The practical suggestions for turning atomic power to good were not slow in coming. As *Newsweek* observed, 'even the most conservative scientists and industrialists [are] willing to outline a civilization which would make the comic-strip prophecies of Buck Rogers look obsolete'. Physicist Alvin Weinberg, later to be director of Oak Ridge National Laboratory, told the Senate in December 1945 that atomic power 'can fertilize and enrich a region as well as devastate it. It can widen man's horizons as well as force him back into the cave.' Among the ideas mooted in the first few weeks of what was already being called the Atomic Age were cars, airplanes, and trains with engines powered by tiny atomic explosions; individual atomic power plants for each house; and personal atomic-powered heaters and air conditioners that could be wired into clothes to keep people comfortable in all temperatures.

The National Education Association assured high-school students that, thanks to atomic energy, 'it is unlikely that you or any of your classmates will die prematurely of cancer or heart disease, or from any contagious diseases, or from any other human ills that afflict us now'. John O'Neill, science writer for the *New York Herald Tribune*, suggested setting off a series of atomic bombs in the Arctic so as to melt the polar ice caps and give the world a 'moister, warmer climate'. The idea received support from World War One flying ace and then-president of Eastern Airlines, Eddie Rickenbacker, and others. O'Neill also predicted that all major American cities would be linked by giant 'vacuum tubes' through which atomic trains would speed at up to 10,000 miles an hour, travelling from Boston to New York, for example, in only ten minutes. Another science writer, David Dietz of the Scripps-Howard newspaper chain, published his book called *Atomic Energy in the Coming Era*, in which he stated that atomic 'artificial suns' would end bad weather: 'No baseball game will be called off on account of rain in the Era of Atomic Energy. No airplane will bypass an airport because of fog . . . A traffic jam due to icy streets will be unknown in the Era of Atomic Energy.' In short, Dietz confidently predicted, 'Universal and perpetual peace will reign in the Era of Atomic Energy.'

The US government was as anxious as anyone to enjoy the fruits of this new era. It was not, however, to be an era of private enterprise. All private ownership of nuclear materials was outlawed, and a new agency, the Atomic Energy Commission (AEC), was created. The commission's public emphasis was on the many utopian possibilities of atomic development, but for many years the commission's main work was the development of new weapons.

The Atomic Energy Commission came into being on 1 August 1946. Twenty days later, Lauriston Taylor decided it was time to revive the United States Advisory Committee on X-Ray and Radium Protection, which had been dormant during the war. The committee faced a new world. Weapons-building, not medicine, was now the major practical application of ionizing radiation. The medical profession had established the committee to provide itself with advice on radiation protection. Now there was a new club of radiation users – the federal government and the private companies who contracted to run its atomic facilities. These new users might want to choose their own safety advisers. Even before the war had ended, various government agencies, including the National Research Council, the Public Health Service, the Army, Navy, and Air Force, had established, or contemplated establishing, their own radiation protection groups. Taylor's committee moved fast to ensure that it would be recognized as the main radiation-protection body. It changed its name to the more official-sounding National Council on Radiation Protection (NCRP). It expanded to include representatives of organizations with an interest in radiation. The new NCRP consisted of eight representatives of medical societies, two of X-ray manufacturers, and nine of government agencies, including the Army, Navy, Air Force, National Bureau of Standards, Public Health Service, and the Atomic Energy Commission. It had been suggested that insurance companies also be represented on the committee, but this idea 'did not find any enthusiastic response', according to the committee's records. Four of the members, including the chairman, Lauriston Taylor, had served on the committee since its founding in 1929. A five-man executive committee, headed by Taylor, dealt with day-to-day problems.

There was a pressing need for new protection standards to carry government, industry, and medicine into the postwar era. The old tolerance limit for X-rays was widely believed to be too high, and there were no standards for the many newly discovered radioisotopes which were chiefly dangerous as internal emitters. The committee recruited outside experts on to technical subcommittees to deal with these subjects.

Many different groups asked the NCRP for advice in its early days. The first request came from a strange quarter. In December 1946, National Selected Morticians, Inc. and the Association of Embalmers asked the committee to prepare a handbook for embalmers who had to handle radiation victims. The report, 'Safe Handling of Cadavers Containing Radioactive Isotopes', was ready in 1953. In 1948 the insurance industry asked for information on the 'possibility of accidental atomic explosions or accidental release of radioactive materials, [and] the probable effects, methods and efficiency of decontamination' so that companies could offer atomic insurance policies.

Feeling that Congress should deal with the issue of nuclear insurance, the NCRP decided not to provide the information. Also in 1948 the NCRP prepared a secret report for the Armed Forces Special Weapons Group on what dose of atomic bomb radiation would be lethal, and how long troops could work while exposed to various levels of radiation before being 'rendered ineffective'.

Towards the end of 1948, Donald Straus, executive secretary of the President's Commission on Labor Relations in Atomic Energy Installations, contacted the committee. The AEC and its industrial contractors wanted to know what special benefits should be extended to workers exposed to radiation hazards. As Straus explained, 'During the war, the extent of the hazards encountered in [Manhattan Project] jobs was unknown. For this reason the government was extremely lenient in allowing special benefits. Such benefit provisions included vacations, sick leave, health insurance and in some cases even a wage differential . . . Special benefits arc still allowed and there is official acknowledgement that extra hazards still exist in certain experimental phases of the work . . . Throughout AEC installations, there are instances of extra benefits being paid to take care of both the known and unknown contingencies . . . These considerations affect collective bargaining . . .'

After consulting with other members of the committee, Taylor told Straus that no special benefits were required for radiation workers, because adequate safety precautions made all radiation work 'nonhazardous'. Taylor explained to Straus: 'I see no alternative but to assume that the operation is safe until it is proven to be unsafe. It is recognized that in order to demonstrate an unsafe condition you may have to sacrifice someone. This docs not seem fair on one hand, and yet I see no alternative. You certainly cannot penalize research and industry merely on the suspicion of someone who doesn't know, by assuming that all installations are unsafe until proven safe. I think that the worker should expect to take his share of the risk involved in such a philosophy . . . I do not think it is feasible to permit unlimited speculation as to unknown hazards.'

By far the most pressing and persistent queries to the committee came from the AEC. Right from the start, according to Taylor, the commission 'was putting considerable pressure on the NCRP to make some kind of pronouncement relative to the permissible dose for radiation workers. They were aware of the fact that the committee was considering lowering the values and naturally this was of great concern to them.' The AEC offered to confer a semi-official status on the committee if it would publish its results quickly, but the offer was declined. Nevertheless, the AEC continued to press the NCRP to speed things up. Commission officials promised the poverty-struck committee financial aid 'after we have demonstrated that we could do something for them,' said Taylor.

The NCRP was in a delicate position. Half its members were government employees. Much of the information it needed was held in secret by the government. The government, particularly the AEC, covered some of the NCRP's administrative and travel costs. The government, particularly the AEC, was to be the chief user of the NCRP's recommendations. The committee was anxious not to be seen as a mere puppet of the AEC, which by 1948 had become a formidable political and economic power, with land holdings larger than Rhode Island. The AEC was equally chary of the NCRP. It needed the backing of a respected and independent scientific body in setting protection standards for its employees, but over-emphasis on safety could impede the development of atomic weapons.

The NCRP had eight subcommittees looking at everything from waste disposal to monitoring methods, but the two most important, as far as the AEC was concerned, were Subcommittee One on the external radiation limit, headed by Gioacchino Failla, and Subcommittee Two on internal radiation limits, headed by Karl Morgan. By the end of 1947 Failla's group had reached agreement on its main recommendation, which was to cut the existing limit for X-rays and gamma rays in half – from 0.1 rem to 0.05 rem per day. The new limit, called the maximum permissible dose, was expressed as a weekly figure – 0.3 rem per week – which gave employers more flexibility in scheduling work assignments than a daily limit.

The reduction was prompted partly by a lack of confidence in the evidence upon which the original tolerance dose was based, and partly by a growing body of evidence that radiation could damage reproductive cells. Experiments with fruit flies indicated that even a tiny dose of radiation could result in the production of mutant offspring. There was as yet no experimental data on radiation-induced mutations in humans, but the NCRP felt it could not afford to wait.

The genetic question put the NCRP in something of a quandary. Up to this point, it had set radiation exposure limits at levels that would prevent 'appreciable' injuries to workers. But from a genetic point of view there was no safe dose, because even the smallest amount of radiation could damage the reproductive cells. However, the committee did not feel it could realistically recommend that radiation workers avoid all exposures. The NCRP was convinced that some exposure had to be permitted, but deciding where to draw the line was a matter of judgement, not science. Ultimately, the committee settled on a figure that the nascent nuclear industry would accept. 'We found out from the atomic energy industry that they didn't care [if we lowered the limit to 0.3 rem per week],' explained Lauriston Taylor. 'It wouldn't interfere with their operations, so we lowered it.'

Failla's group had agreed on the new maximum permissible dose by the end of 1947, but more than six years passed before the report containing this advice was published. The delay was caused by controversy about other recommendations in the report. Failla's main critic was Karl Morgan, who besides heading Subcommittee Two was chief health physicist at Oak Ridge National Laboratory. Among the recommendations that Morgan objected to was one which allowed workers over 45 years of age to receive 0.6 rem per week, twice as much as younger workers. The rationale for this was that older people could stand more radiation injury. They were past their childbearing years, so any genetic damage they sustained would not be passed on to future generations; and their life expectancy was shorter than the latency period of many radiation-induced diseases. Morgan disagreed. 'It seems to me,' he wrote in a letter to Taylor, 'from the standpoint of cataract production, blood changes, and cancer incidence that it is just this group that have reached 45 and already accumulated a lifetime exposure that we are most interested in protecting.' Morgan argued against a dual exposure system on practical grounds as well, saying that it would be difficult to administer.

Morgan also took exception to another provision, one which allowed workers to receive, in addition to the regular permissible exposure of 0.3 rem per week, an extra 25 rems either in a single accident or a series of accidents while they were 45 years of age or younger. Workers over 45 could receive another 25-rem 'accidental' dose on top of the first. Morgan was concerned that giving advance permission for 25- or 50-rem accidental exposures would encourage carelessness, and give employers and scientists the idea that, accidents or not, they ought to 'use up' these permitted exposures. He cited evidence that a 25-rem exposure to a woman in the first few weeks of pregnancy (when many women might be unaware that they were pregnant) 'can lead to a rather high incidence of abnormalities'.

The philosophical differences between the two subcommittee chairmen were so great that they could not be resolved directly. Finally Taylor got Failla and Morgan to give him the final say. In Failla's report, published in 1954, the weekly limit for workers over 45 years was set at 0.6 rem, twice as much radiation as younger workers were allowed to receive. The idea of an emergency exposure allowance was also retained – a one-time extra exposure of 25 rems for workers of all ages. It was understood, however, that there could be no restrictions on emergency exposures to military personnel. 'Radiation risks would have to be considered as a part of the many other risks that accrue to military personnel,' said Taylor.

Morgan's group, dealing with internal emitters, had a daunting task. Its job was to set a limit on radioactive substances inside the body. Failla's group had dealt with external exposures, which tend to irradiate the whole body evenly. By contrast, radionuclides inside the body irradiate mainly their 'target' organ, which varies from radionuclide to radionuclide.

The first thing the subcommittee did was to set dose limits for each of the major body organs – the gonads, red bone-marrow, thyroid, lung, bone surfaces, and, for women, the breast. Different limits were set for each organ, according to its sensitivity to radiation. If each organ were irradiated to its limit, there would be approximately the same chance of developing cancer as if the whole body had received the allowable dose of external exposure. There was no mechanism for calculating a person's total exposure if several organs each received internal doses, or if a person received both internal and external exposures. The underlying assumption was that a person working with radiation would either suffer whole-body exposure or internal exposure to only one organ.

The only practical way to make sure that no organ was over-exposed was to limit the amount of each radionuclide in the working environment. Morgan's group therefore had to decide how much of each radionuclide would be allowed in air and water. It was a formidable undertaking which required analysing not only the radioactivity of each radioisotope, but also its behaviour in the body. Is it quickly expelled, or does it pass slowly through the system? Does it concentrate in the bones, the liver, the bone marrow? The answers to these questions vary according to the radioactive material's chemical properties and whether it was swallowed, inhaled, or injected into the body. Morgan's group based its calculations on a theoretical 'standard man', who lives to be 70 years old, who weighs 154 pounds, 66 pounds of which is muscle, whose water intake is 74 fluid ounces a day, who sweats 16 fluid ounces of sweat a day, and so forth. Though 'reference man' was a convenient concept for the subcommittee, it overlooked the fact that individuals vary greatly in their susceptibility to radiation, according to age, sex, and other, unknown, factors. A few per cent may be five times more sensitive to radiation than the 'average' person.

The information about radioisotopes was scarce and often conflicting. Much of it was also classified due to its military applications, but close ties with the AEC enabled the subcommittee to get most of the data it needed. In 1948, Herbert Parker, a prominent health-physicist and member of the NCRP subcommittee, noticed that every single member of the subcommittee was connected with the AEC. 'From the public relations angle,' he warned, 'this appears to leave the Commission in its former vulnerable position of writing its own ticket.' But Parker's suggestion that outsiders be brought into the group was rejected. Instead, according to Taylor, 'as time went on, more individuals connected with the Commission generally, or with Oak Ridge in particular, were added. This had certain advantages in that Morgan could influence the amount of effort that they would put into their [sub]committee work.'

Subcommittee Two originally planned to recommend maximum permissible concentrations (MPC) in air, water, and the human body, for 20 radioisotopes, but the number eventually grew to 96. In fact, fretted Taylor, the members were 'never able to find a cut-off point in their considerations'. The subcommittee's work dragged on. 'We cannot be forever postponing action hoping for better things tomorrow,' Taylor complained. In 1951 the NCRP's executive committee summarily ended Subcommittee Two's deliberations and insisted that its report on internal emitters be prepared for publication. But another two years elapsed while the NCRP argued about some of the statements in the subcommittee's report.

Leading the attack on Morgan's report was Failla. He took exception to the group's decision to suggest that MPCs should be divided by a 'safety factor of ten' in cases where people might be exposed for long periods of time, 30 years or more. The subcommittee's reason for suggesting this was lack of data about the effects of prolonged exposures at MPC levels. They were concerned about giving their recommendations, in Taylor's words, 'a ring of authority which had no basis in fact'. But Failla argued that too many cautions and qualifications would undermine public confidence in the report and cause people to disregard it entirely. He thought 'all that was necessary was to emphasize that these were interim values subject to change at any time'. Taylor intervened to settle the dispute. The final report, published in 1953, does not recommend lower doses for people likely to suffer prolonged exposures. Instead it suggests that 'large permanent [nuclear] installations' be designed so that they could meet more stringent standards if they were introduced sometime in the future.

In 1953, the NCRP's standards were adopted wholesale by the international radiation protection commission which, largely due to Lauriston Taylor's efforts, was resuscitated after the war under a new name, the International Commission on Radiological Protection (ICRP). Only two members of the original committee had survived the war, Taylor himself, (who was also chairman of the NCRP) and Rolf Sievert of Sweden. The reformed group had nine members – three from Britain, two from the USA, and one each from Canada, France, Germany, and Sweden. It also established technical subcommittees modelled on those of the NCRP. Failla was chosen to head the external dose subcommittee, and Morgan to head the subcommittee dealing with internal emitters – the same ones they chaired for the NCRP. Britain, Canada, and the United States – especially the latter – dominated the organization. Despite its international membership, the ICRP, in its early days, was little more than the overseas branch of the NCRP.

Legally, the NCRP had the best of both worlds. Everyone treated its recommendations as if they came from an official body, yet because the recommendations were not incorporated into rigid laws or regulations, the committee could change them at any time. This was important, said Taylor, because the committee 'was quite aware that it might be some years before our knowledge and philosophy of radiation protection techniques would be sufficiently clarified to warrant the introduction of such rigidity'. Another reason for the NCRP's aversion to legislation was the likelihood that other groups would succeed in influencing the rule-makers. Such influences would necessarily be negative, since, as Taylor confidently asserted, 'The only qualified protection experts in the country are already members of this committee.'

The AEC, on the other hand, wanted standards to be legally binding so that they could not be altered at the NCRP's whim. The AEC wanted protection against sudden changes that might require expensive alterations to the design and operation of nuclear plants. When it became clear, in the early 1950s, that both the federal government and some states were going to formalize radiation protection standards, the NCRP cooperated, in order to ensure that the standards were based upon its recommendations. The first federal radiation regulations became effective in 1957.

The quasi-legal status afforded to the committee's recommendations concerned some members. In a letter to Taylor in late 1949, Robert Stone, the former chief of the Manhattan Project's Health Division, now returned to his civilian job as head of radiology at the University of California's San Francisco Medical School, remarked that, 'since these books [of NCRP recommendations] get a quasi-legal status, they might put some of us in an embarrassing situation when we do not follow to the letter the recommendations that are made'. Stone said that he had 'not followed the rules regarding pre-employment, yearly, or discharge examinations with regard to personnel in my department' and asked whether such behaviour might open him and other members up to legal action. From time to time over the years, other members again raised the issue, notably in 1970 when Herbert Parker asked for a legal opinion on the liability of individuals participating in NCRP affairs. Parker was worried about damage suits that might be brought by persons claiming that the committee's recommendations had not protected them from injury. The committee was assured that there was little chance of such a lawsuit succeeding. This response did not mollify Parker. He asked the committee to look into providing legal insurance for its members, but no further action was taken.

The NCRP did not have a formal philosophy of radiation protection, but one emerges from the assumptions and decisions the committee made in reaching its conclusions. One of the key points was the 'no-threshold' theory, which was officially accepted by the NCRP in 1948. This theory holds that there is no absolutely safe level of radiation exposure. It was this view that caused the NCRP to follow the Manhattan Project's lead in substituting the phrase 'maximum permissible dose' (and, for internal emitters, 'maximum permissible concentration') for the phrase 'tolerance dose', to better express the concept that no dose was absolutely harmless. Maximum permissible dose was defined as an amount of radiation which, 'in the light of present knowledge, is not expected to cause appreciable bodily injury to a person at any time during his lifetime'. The phrase 'appreciable bodily injury' was defined as any effect or injury that 'the average person' would regard as being objectionable, or that 'competent medical authorities' would regard as being deleterious to an individual's health and well-being.

The committee did not oppose all avoidable radiation exposure. For one thing, it believed that doing so would encourage the general public in a 'psychologically dangerous' fear of radiation – one that could lead to panic in the event of an atomic explosion. More to the point, the committee was convinced that – at the exposure limits it set – the risks of radiation exposure were acceptable because they were outweighed by the benefits to be gained from the use of radiation. This raised a number of interesting questions. What are the political, social, and economic benefits that arise from the construction of atomic bombs, the development of nuclear power, or other uses of ionizing radiation? Who is qualified to evaluate them? And who should determine whether or not a given risk is acceptable, especially when the risks and benefits are not equally distributed among the population, or when they are not voluntarily or knowingly assumed?

CHAPTER NINE

Uranium Fever

In the decade after the war, uranium fever swept the United States. In 1953 alone, Americans bought 35,000 Geiger counters. Finding uranium became a patriotic duty, as expressed by Gordon Dean, chairman of the Atomic Energy Commission from 1950 to 1953:

> The security of the free world may depend on such a simple thing as people keeping their eyes open. Every American oil man looking for 'black gold' in a foreign jungle is derelict in his duty to his country if he hasn't at least mastered the basic information on the geology of uranium. And the same applies to every mountain climber, every big game hunter, and, for that matter, every butterfly catcher.

The prospectors did their work well. Deposits worth exploiting were discovered in Australia, South Africa, France, Niger, and Gabon, and secured for the free world. Uranium, it became clear, was not as rare as had been supposed. Nonetheless, the United States preferred the security of a domestic supply. Although one pre-war American dictionary defined uranium as 'a worthless metal, not found in the United States', uranium-bearing ores had been mined in Colorado and Utah in the early years of the twentieth century for the minute quantities of radium that could be extracted from them. The Uraven mine, south of Grand Junction, Colorado, is reputed to have mined radium for Madame Curie, although most of her supplies came from Bohemia and later Shinkolobwe in the Belgian Congo. To encourage prospectors, the AEC opened offices in Colorado, New Mexico, and Utah, published advice on how to prospect for uranium, set a guaranteed price for uranium, offered development loans, and announced a $10,000 discovery bonus for new high-grade deposits.

The focus of the excitement was the Colorado Plateau, a 120,000-square-mile region where Colorado, Utah, Arizona, and New Mexico come together. The Colorado Plateau is the site of the 25,000-square-mile Navajo Reservation, and those of several smaller Indian nations. Ironically, the apparently worthless land that the US government had ceded to the Indians turned out to contain a large percentage of the country's uranium deposits. Indians were excellent prospectors because they knew the land so well. Shown samples of the numerous types of uranium-bearing ore, many were able to retrace their steps to exactly where, even years before, they had seen similar rocks. One of the first people to make a big strike was a Navajo shepherd named Paddy Martinez. His discovery of the Haystack mine near the small town of Grants, New Mexico, which soon gave itself the title of 'The Uranium Capital of the World', led to the realization that the area contained one of the world's largest uranium deposits.

Feeding on the uranium boom, magazines published articles with titles like 'Get Rich From That Miracle Atom'. Pamphlet writers were equally enthusiastic: 'Uranium – creating new men of wealth, uraniumaires! Among the uraniumaires are yesterday's stock clerks, students, scientists, dentists, engineers, and brokers. Tomorrow it could be you.' Thousands of men, and a few women, left their families behind and staked everything on a risky and exhausting search through the spectacular but rough mesa country. More cautious types became 'rock hounds', scouring roadsides on weekends with mail-order catalogue Geiger counters. Families took counters along on Sunday outings, just in case. Real stay-at-homes could buy shares in the undeveloped mines. Advertised on radio, in newspapers, and even by men wearing sandwich boards, the shares were sold at gas stations, bars, and motels, usually for a penny apiece. Some restaurants and shops gave uranium stocks away with every purchase of a hamburger or a tube of toothpaste. In the month of May 1954, 30 million uranium mine shares were sold in Salt Lake City alone. Prospectors who made a strike became obsessed with raising the money to actually bring the ore out. According to Raymond Taylor, one of the period's many paper millionaires, a common sign in bars and diners all over the plateau was 'No Talk Under $1,000,000'. Some fortunes stayed underground for lack of development capital, but there were also spectacular successes. Many of the winners lived in Moab, a sleepy valley town in southern Utah that soon claimed to have more airplanes per capita than anywhere else in the world. Moab housewives flew to Grand Junction, Colorado, to buy superior groceries. One Moab 'uraniumaire' solved the problem of the area's poor television reception by putting a TV set in his plane and having his pilot take him up every afternoon for the Mickey Mouse Club Show.

By the mid-1950s, hundreds of uranium mines – many of them small operations with fewer than six employees – were operating on the Colorado Plateau. In the early days, the solitary prospector was a common sight, and loners discovered many of the biggest deposits, but big companies such as Kerr McGee, Anaconda, Union Carbide, United Nuclear, and Homestake Mining, increasingly dominated the industry, their paths smoothed by favourable treatment from the AEC. By the end of the 1950s, when the government ended the first uranium boom by announcing that it would not buy uranium from new deposits, Kerr McGee had rights to about one-quarter of the country's known uranium reserves. Big or small, one thing all mines had in common was the risks the miners were taking, and their ignorance of those risks.

There are three naturally occurring isotopes of uranium, and all of them are radioactive. The most common isotope, uranium-238, goes through seventeen changes before stabilizing as a non-radioactive form of lead, lead-206. Any uranium deposit will contain traces of uranium's radioactive decay products, including thorium, radium, polonium, lead, and radon, a radioactive gas. Radon decays into a series of short-lived radioactive substances, known as radon daughters, all of which emit alpha particles. In undisturbed deposits, radon and its daughters are trapped in the rocks. When mining begins, the gas is released, and the radon daughters attach themselves to tiny particles of dust and smoke in the enclosed space. Because uranium is present in some degree in most mineral deposits, even non-uranium mines may have high concentrations of these alpha-emitting substances. When they are inhaled, the radioactive particles settle on the lungs and irradiate them intensely at close range for many years, even after death. Even a single alpha particle hitting a single cell can set in motion the process by which cancer develops, though there may be no symptoms for ten, twenty, or more years.

In addition to producing alpha particles which can irradiate the body from within, uranium and its decay products also give off gamma rays. The average uranium miner gets an external dose of 200 millirems (a millirem is 0.001 rem) per year from gamma radiation, which is among the highest exposures for any radiation occupation.

The miners of the uranium rush had never heard of radon daughters, alpha particles, or gamma rays, but they were intimately acquainted with them. They worked in tunnels filled with radioactive smoke created by the drills and dynamite used to loosen the ore. They ate their lunches inside the dusty, unventilated mines, drank contaminated water from underground springs, and went home in their radioactive, dust-covered workclothes to hug their wives and children. They were not told of the risks of radiation, not monitored for radiation exposure, not given medical check-ups or protective equipment.

Speaking in 1979 to a Senate committee, former miner George Kelly described his work for the Alvin Burwell Mining Company on the Navajo Reservation at Shiprock, New Mexico. 'Inside the interior of the mine was a nasty area, smoky, especially after the dynamite explodes. We run out of the mine and spend five minutes here and there and were chased back in to remove the dirt by hand in little train carts . . . The water inside the mine was used as drinking water, no air ventilators, however. The air ventilators were used only when the mine inspectors came and after the mine inspectors leave, the air ventilators were shut off . . . What really disappointed me were the mine inspectors when they arrive, only one would stick his head in for about five minutes and run to the outside again. Other than that, sometimes not one of them would proceed into the mine because of the smoke.'

Fifteen, twenty, or even thirty years after leaving the mines, many of these men developed lung cancer and other respiratory diseases, such as pulmonary emphysema, fibrosis, and chronic bronchitis. The average age of the miners who die of lung cancer is 46. According to the US Public Health Service, the death rate from lung cancer is five times greater among uranium miners than in the population as a whole. The death rate from non-malignant respiratory diseases is more than three times greater for uranium miners than for the general public.

'This thing comes back every day, every night. It's been fourteen years, and I always dream about it,' said Phillip Harrison, whose father, like many Navajos, was a uranium miner. 'I remember in one dream saying "Dad, we'll take care of you this time." He was a miner, with a miner's build – five foot nine, husky, and by the time he died, you know, we put him away at 90 pounds. Just skin and bones.'

Phillip Harrison senior died in 1971, 43 years old. His son, Phillip, who himself worked in a uranium mine during school vacations, now works for the Environmental Health Division of the Public Health

Service in Shiprock, New Mexico. Shiprock is now best known for its 2.5 million tons of uranium tailings, the radioactive residue of the town's once-thriving uranium mill, which processed uranium from the hundreds of mines on the reservation. Whoever arranges these matters in Shiprock has an interesting sense of humour: Harrison's office backs directly on to Shiprock's biggest environmental health problem, the 72-acre tailings pile.

As a child, Phillip Harrison lived with his parents, two brothers and three sisters on the Navajo Reservation, which covers the northeast corner of Arizona, and spills over into New Mexico and Utah. Their home, and the Kerr McGee mine Phillip Harrison senior worked in for 16 years, was near the small settlement of Cove (population about thirty families) in the Red Valley, a sparsely-populated area rich in uranium and in the problems that stem from mining it. In a 1974 report on energy policy the National Academy of Sciences recommended that in view of the strategic importance of uranium and the intractable environmental and health hazards it poses, the Red Valley and other uranium centres be declared 'National Sacrifice Areas' and abandoned to uranium mining and processing.

'I have seen too many funerals,' Harrison told me. 'I've been to most of the families, and they are having a hell of a time getting themselves straightened back up. You know, after dealing with your father for maybe his last two years, being with him to the day he dies . . .

'My father, the last year of his life he had greatly suffered; he had really suffered daily. We gave him pain pills, but the pain just started mounting and pretty soon the pain pills weren't enough. They started shooting him with needles, and the needles didn't stop the pain. I think they die mostly from pain. I just lost an uncle here; his funeral was April 9th. He was a miner, too. And watching him, the pain overcame the shots, so there was nothing they could do. We had another funeral that week, and we have another one this week. Three different people, all miners, all lived around Cove.

'None of those people ever said anything about having a safety meeting or any kind of orientation that would tell them about the risks of uranium mining,' Harrison recalled of his father's generation of miners. 'And when I went to work [in 1969], I was never told anything inside the mine would be hazardous to my health later. It really surprised us to find out after so many years that it would turn out like this, that it would kill a lot of people. They said nothing about radiation or safety, things like that. We had no idea at all.'

No one knows for certain how many men have worked in underground uranium mines, who they are, or how many have contracted lung cancer or other respiratory diseases. Harrison now keeps a record of the ex-miners with health problems, as a first step towards getting compensation and help for the mining families, an uphill struggle. Compensation for occupational injury is usually a state matter, though there are exceptions. For instance, the public concern aroused in the 1960s by the suffering of the Appalachian coal miners led to a federal compensation and treatment programme for black lung disease. No such programme exists for uranium miners. The state compensation laws are not designed to deal with the long latency period and the invisible progress of radiation-induced damage. They contain statutes of limitation that require proof of injury within a few years, at most, of leaving a job. Many also require evidence of working conditions that state and company records do not provide. In the not-uncommon case that a miner has worked in several states, each one may deny compensation by claiming that the damage occurred in the other.

Former Secretary of the Interior, Stewart Udall, who has been trying since 1979 to get compensation for the families of dead and disabled Navajo uranium miners, has said of the compensation laws, 'I have seen a lot of buckpassing in the government in my day, and I must say this is the most outrageous example that I have ever seen.'

The link between uranium mining and lung cancer was well known when the post-war uranium boom began. It was, in fact, 'the most well documented case of an occupational disease that existed', according to Duncan Holaday, former director of industrial hygiene at the US Public Health Service. The evidence came from mining communities in the Erzgebirge, the mountain range that divides Bohemia and Saxony (now Czechoslovakia and East Germany). Mining of pitchblende, the area's uranium-rich ore, for silver and other precious metals began in the early 1400s. Joachimsthal on the Bohemian side provided silver so pure than the coin made from it, the Joachimsthaler, thaler for short, was used throughout central Europe. The thaler eventually gave its name to the American dollar.

The Bohemian village of Jachymov, where mining began in 1516, had as its town physician George Bauer, better known by his pen name of Agricola. Bauer noticed that many Erzgebirge miners died prematurely of respiratory disease. 'Women are found who have married seven husbands, all of whom this terrible consumption has carried off to a premature death,' he wrote in 1546. Agricola blamed the problem on the ore itself, whose dust, he wrote in *De Re Metallica*, his classic treatise on mining, 'has corrosive qualities, it eats away the lungs and implants consumption in the body'. His contemporary, Paracelsus, also attributed the miners' illnesses to their inhaling metallic gases 'which settle on the lung'. The miners believed that their 'mountain sickness' was the work of spiteful underground dwarfs. But, along with Agricola, they thought it important to ventilate the mine shafts and take other steps to reduce their inhalation of the ore dust. Miners' wives became adept lace-makers, fashioning fine veils for their husbands to wear underground. Their precautions were sensible, but the evil dwarfs were not so easily foiled.

In 1879 Martin Klaproth, a German chemist, obtained a sample of pitchblende from the Erzgebirge, and discovered that it was a new element. He named it uranium, in honour of the planet Uranus, discovered eight years before. Also in 1879, two German doctors identified the 'mountain sickness' of the pitchblende miners as lung cancer. In 1913 another study showed that 40 per cent of the miners in one Erzgebirge village who died between 1875 and 1912 were killed by lung cancer. In 1932 two researchers reported that 53 per cent of the deaths among pitchblende miners in Joachimsthal were due to lung cancer; they concluded that radiation in the air of the mines was the most probable cause. In a follow-up study seven years later, the Joachimsthal miners were found to be twenty times more likely to die of lung cancer than non-miners in Vienna. The Joachimsthal miners also suffered from a higher rate of skin cancer. By 1942 lung cancer among pitchblende miners was being cited as the classic example of cancers associated with exposure to radioactive substances. In the 1940s the French were installing ventilation systems in their uranium mines, as the Czechoslovaks had done in the 1930s.

In the United States, however, uranium miners had no protection, legal or physical, against the deadly radon daughters. Under the 1946 Atomic Energy Act, the Atomic Energy Commission controlled the uranium industry. No one else was permitted to own uranium. All uranium that was mined had to be sold to the AEC. The commission even operated some of the mines directly, but it declared itself not responsible for protecting the health of the miners. Its 1951 pamphlet, *Prospecting for Uranium*, makes no reference to radiation, except to say that 'the radioactivity contained in rocks is not dangerous to humans unless such rocks are held in close contact with the skin for very long periods of time'.

Working conditions in mines were traditionally a state responsibility, and the commission wanted to keep it that way. But with uranium, things were not so simple. Many of the mines were on Indian lands, where states have no jurisdiction. The secrecy and aura of national security that surrounded the federally controlled uranium industry made it difficult for state authorities to know what their powers and duties were. State inspectors were often discouraged or even barred from entering the mines on the grounds that they were hampering work essential to the nation's survival. In any case, for years the states paid no attention to radiation protection, which they did not have the scientific expertise to regulate or the equipment to monitor. In 1948 only two laboratories in the entire country could measure alpha activity, the federal government's Bureau of Standards in Washington, DC and the Massachusetts Institute of Technology.

The standard, adopted in 1941, of ten picocuries (a curie is a measure of radioactivity; a picocurie is one trillionth of a curie) per litre 'in plant, laboratory, or office air', was deemed not to apply to mines. In 1949, the Atomic Energy Commission's New York Operations Office, which was responsible for procuring uranium, tried to get the radon standard and other health protection clauses written into the commission's contracts with uranium mining companies. 'We took the position . . . that since we were the only customer, we should see to it that the standards that already existed could be met,' Merril Eisenbud, ex-director of the AEC's New York office, later said. 'We would have written that standard into the contracts, but we were not permitted to do so.' Shortly afterwards the commission stripped the New York office of its procurement responsibilities.

In 1953, the NCRP recommended that the 10-picocurie-per-litre limit be extended to uranium mines. But, at a conference later the same year, the United States, Britain, and Canada agreed to a limit ten times higher, 100 picocuries per litre. In 1959, with the blessing of the US Public Health Service, the more lenient limit, 100 picocuries, became the US standard. One hundred picocuries per litre of air was dubbed a working level (WL).

Although there was little public concern in the 1950s and early 1960s about the risks that uranium miners might be running, some scientists were pressing the AEC to act to protect miners. One problem was the lack of information about the amount of radiation miners were exposed to. Neither the mining companies nor any state or federal agencies monitored the mines regularly until the mid-1960s or later. In 1950 the US Public Health Service began a survey of uranium mines. It found astonishingly high levels of alpha exposure throughout the industry. In 1952, for instance, the average uranium mine exposed its workers to 8 WLs of alpha activity (mostly radon and radon daughters). Forty-five per cent of uranium miners were exposed to more than 10 WLs. According to a 1981 US government review of uranium regulations, the one WL limit, adopted in 1959, 'was effective in reducing average industry-wide WL values from 7 WL in 1957 to 2.1 WL in 1966'. In other words, seven years after the one WL limit was adopted, the average US uranium mine was more than two times over the legal limit.

In 1960 Dr Joseph Wagoner, an epidemiologist, was assigned to analyse data that the Public Health Service had collected on 3,400 white underground uranium miners. This was the first study of the American uranium industry and the results were shocking. Dr Brian MacMahon of Harvard, who reviewed the Health Service study in 1962, concluded that the results 'give no comfort to those who are concerned that the problem in American mines may become a disaster . . . as great or even greater than the European experience'.

'It was in 1962 that we first showed in our own miners an increased risk of lung cancer,' Wagoner, speaking in 1980, recalled. 'In 1964 we first showed that the long-term underground miners have a ten-fold increased risk of lung cancer. And it was in 1965 that we found lung cancer risks increase with greater radiation exposure and that cigarette smoking was not a factor in this dose–response relationship.'

In the summer of 1967, with the evidence mounting that hundreds or thousands of uranium miners would die prematurely of lung cancer, Congress's Joint Committee on Atomic Energy held hearings on the epidemic of lung cancer in uranium miners. The AEC and the Federal Radiation Council (FRC), which was responsible for recommending radiation exposure standards, opposed any change in the radon standards, arguing that the industry could not bear the cost of improving ventilation in the mines. Dr Paul Tomkins, the director of the Radiation Council, told the committee that 'the primary objective of the FRC is to make recommendations which represent a reasonable balance between biological risk and the impact on uranium mining'. After the hearings the council recommended that records be kept of how much radiation each miner received, but not that the exposure limit be lowered.

Meanwhile, Merril Eisenbud had left the AEC and become director of New York University Medical Center's Laboratory for Environmental Studies. With a graduate student he developed an instrument that could 'read' each miner's past radon exposure, by measuring the body's burden of lead–210, a long-lived decay product of radon. They used the device on 100 miners in Denver and Utah and found it an effective way of identifying those at greatest risk of contracting lung cancer or radiation-related illnesses. Eisenbud urged the federal government to adopt the instrument, but both the AEC and the Health Service said they did not have enough money to carry out a screening programme. 'It is hard to believe that, because we were talking about a very small amount of money,' Eisenbud told a joint Congressional hearing on the health impact of low-level radiation in 1979.

Throughout the 1960s, the Public Health Service's study of former uranium miners continued, with disturbing results. A Public Health Service administrator, Charles Johnson, announced in 1969 that 'of the 6,000 men who have been uranium miners, an estimated 600 to 1,100 will die of lung cancer within the next twenty years because of radiation exposure on the job'.

In 1971, after several years of debate and public pressure, and the dissolution of the Federal Radiation Council, its successor, the Environmental Protection Agency, ruled that the average concentration of alpha emitters in uranium mines must not exceed one-third of a working level. A new unit, the 'working level month', was also introduced. One 'working level month' is exposure to one working level (100 picocuries per litre of air) for a period of one month. An average concentration of 1 WL for one year would give an annual exposure of 12 WLMs. One-third of a WL for one year gives an annual exposure of 4 WLMs.

'From twelve to four, no one thought they could ever do that. It took a real overhaul in ventilation, but we've never had anyone over-exposed.' John Parker, the manager of Homestake Mining's uranium operation, about five miles outside Grants, is now 'quite happy' with the limit of 4 WLMs. 'For the last few years we've been sampling every working place, every shift.' In the old days, he freely admits, things were different. 'When there were no regulations, the mines were rat-holes.' Even in the 1960s when the limit was 12 WLMs a year, 'I'm quite sure that some workers were exposed to 1,000 WLMs over a lifetime.' In fact, the Public Health Service study showed that some workers had been exposed to more than 3,000 WLMs.

The company does not monitor mine air continuously because 'there's no gadget for that', says Parker. Many workers wear film badges to keep track of gamma exposure, but film badges cannot detect alpha radiation. Instead, each worker must keep a record of how much time he spends in each area of the mine. A computer calculates exposures by comparing each worker's timesheet with the periodic air readings for the areas in which he worked.

The Mining Safety and Health Administration (MSHA) inspect uranium mines, by appointment, once each quarter. State officials generally come on a monthly or weekly basis. But for day-to-day exposure records, the government relies on readings taken by company officials, a reliance that Parker, and the uranium industry as a whole, say is justified. The companies' records back up the assertion of Simon Rippon, a British nuclear physicist and author, that 'in all present-day operations radiation exposure levels to miners are kept well below the internationally accepted limits'. In 1977 uranium companies reported that only 0.2 per cent of all uranium miners had received more than the legal limit of 4 WLMs per year, and that the average exposure was 0.91 WLM, well under the limit.

This assertion was contradicted by a monitoring 'blitz' that MSHA began in 1975. The MSHA operation demolished the image of a self-policing uranium mining industry. A team of federal inspectors paid surprise visits to 20 underground uranium mines, employing more than 1,600 miners. According to the MSHA director Robert Lagather, 'it was necessary to have more than one inspector present at the mine during the inspection. This made it impossible to adjust ventilation from one area of the mine to another to satisfy the inspector while allowing another area to experience increased radiation levels.' The inspectors found that many miners were receiving over the legal limit of 4 WLM. The average miner was exposed to 4.64 WLMs in a year, five times what the companies claimed. In addition, many miners received extraordinarily high doses that did not appear on company records when the mineface was blasted open, filling the mine with large amounts of radioactive dust and debris. 'Radon daughter concentrations,' said the official report, 'which were at 0.2 WL during the morning would jump to as high as 17 WL during the first hour after lunch or

85

during the first hour at the start of the next shift. A miner would need to be exposed to those levels for only 15 minutes a day for 160 days to exceed federal exposure standards.'

As the history of uranium mining in the United States approached the 20- to 30-year latency period of cancer, miners began to see among themselves and their fellow workers the effects of the years spent underground. By 1978, 205 of the 3,400 underground miners in the Public Health Service study had died of lung cancer. The government estimates that 30,000 to 40,000 men have worked in underground uranium mines. If the study is representative, then thousands of these miners have died or will die of lung cancer, and thousands more from emphysema, fibrosis, and other respiratory ailments.

James Aloyisius started at the Kerr McGee's Cove mine, on the Navajo Reservation, in 1957 when he was 27. 'Nobody never did say anything,' he remembers. 'All they did was give you a job. There were people dying then, but we didn't know what it was.' After five years underground, Aloyisius became a foreman and worked above ground for the next 19 years, mostly at Union Carbide's Dovecreek mine in Colorado. He first became aware of a problem in the mines in 1980. 'So many of the people there in Dovecreek were dying with lung cancer. So I started asking what this was. They said it was all the dust in the mines. A lot of people I worked with died.' Aloyisius quit the mines in 1984, convinced that they were unsafe. Asked about improvements over the last two decades, he nodded, 'They done some work on it. There's good ventilation in the main channel, but there's a lot of bad air in the side tunnels.'

The companies don't deny that uranium miners are suffering very high rates of lung cancer, but they say smoking, not working in the mines, is the main cause. Some miners, say Homestake officials, break company rules to smoke underground. 'Did you ever like to sneak a cigarette past your parents?' Homestake's safety manager, Red Souther, asked me. 'Those are the ones that taste best. That's what they do with us. It's an act of rebellion against authority.' Smoking does increase a miner's risk of getting lung cancer, but even non-smoking miners are many times more likely to get lung cancer than non-miners. Nonetheless, miners say that some companies continue to tell them that if they do not smoke, they will not develop lung cancer from exposure to radiation in the mines.

Most of the miners who have suffered so far worked underground before 1971, when the 4-WLM limit was set. Mining companies and most regulators maintain that in the past decade tougher standards, and better policing, have turned underground uranium mines into safe places to work. They believe that in a few years this spate of premature deaths, what Joseph Wagoner calls 'this epidemic and public health problem of monumental proportions', will be over. But there is a growing body of evidence from other countries that even 4 WLM is too high for safety.

The 4-WLM limit, like other radiation standards, was chosen almost arbitrarily. Scientists did not and still don't know exactly how radio-activity induces cancer in humans, so setting exposure limits has been largely a matter of guesswork. The experts have developed complicated mathematical equations that they hope express the relationships between certain radiation exposures and the likelihood of cancer, but, as Robley Evans, who originated the first standard for radon, told Congress's Joint Committee on Atomic Energy in 1967, 'some people feel then you might as well guess at the answer instead of guessing at the numbers to put in a complicated formula. It is shorter.'

When it became apparent that the 1 WL limit was too high, regulators looked around for a suitable lower number. In 1959 the International Commission on Radiological Protection backed a limit of 0.3 WL, which over 12 months would result in an exposure of 3.6 working level months. According to Robley Evans, the ICRP set that limit by mistake. Just before the war the radium limit was set at 10 picocuries per litre of air, which is equivalent to 0.1 WL. This 'was through some inadvertence interpreted as relating to continuous exposure for the whole week, 168 hours, instead of 40.' Thus, said Evans, the Commission wrongly thought it would be all right to increase the permissible concentration from 0.1 WL to 0.3.

The Department of Labor had also seized upon the figure of 0.3 WL, partly because officials believed, wrongly, that the Public Health Service had said that the natural background level of radon in the environment was 0.3 WL. The figure of 0.3 WL, or roughly 4 WLMs per year, was bandied about for so long that it became something of an institution.

Regulators set the 4-WLM standard on the assumption that a miner who worked for 30 years and was exposed to a total of 120 WLMs would not have a significantly greater risk of developing lung cancer than a non-miner. Since the 4-WLM standard was adopted, however, studies from Czechoslovakia, Canada, and Sweden have shown excess fatal cancers in miners exposed to only 40 or 50 WLMs. In 1976 the US National Institute for Occupational Safety and Health (NIOSH) concluded that the 4-WLM-per-year standard 'cannot be characterized as safe since the risk of lung cancer would be expected to double after 10–20 years employment'.

On 21 April 1980 the Oil, Chemical, and Atomic Workers Union, whose members include some uranium miners, filed a petition with the Mining Safety and Health Administration, asking that the limit for alpha emitters be cut from 4 WLMs to 0.7 WLM per year. The government promised a 'serious review' of the petition, and asked NIOSH to review the evidence. In June 1985 NIOSH published a report, which, according to Larry Mazzuckelli, who oversaw its completion, concluded 'that the current standard was not sufficiently protective'. In 1986, MSHA proposed that the standard be lowered to roughly 2 WLMs a year, but the agency has not yet made the new standard official.

CHAPTER TEN

Operation Crossroads

Early in 1946 the United States announced that in mid-May a military task-force would conduct a series of atomic tests at Bikini Atoll in the Pacific Ocean. These would be the first atomic explosions since Hiroshima and Nagasaki were destroyed. The test series, named Operation Crossroads, was to consist of three 23-kiloton explosions, Able, Baker, and Charlie, one to be detonated in the air and two underwater. Bikini Atoll, 4,500 miles from the continental United States, is part of the Marshall Islands chain. During the war the Marshalls were occupied by the Japanese; shortly afterwards the United Nations assigned the islands to the protection of the United States. Four months before the tests, the Navy moved the 160 Bikinians one hundred miles away, to Rongerik, another Marshall Islands atoll. A US government newsreel of the time depicts the Bikinians as delighted with the move: 'The islanders are a nomadic group and are well pleased that the Yanks are going to add a little variety to their lives.'

Vice Admiral W. H. P. ('Spike') Blandy was commander of Operation Crossroads; Manhattan Project veteran Stafford Warren was put in charge of radiological safety. Crossroads was designed to provide information on the effects of atomic weapons on naval vessels. The target fleet, eighty-four Japanese, German, and American ships loaded with military equipment, including tanks, planes, clothing, and ammunition, would be anchored in the 200-square-mile shallow lagoon during the blasts. The 'live fleet' of 100 ships, which would be home to most of the 42,000 troops and civilian scientists participating in the tests, would wait outside the test area. Afterwards the target vessels that survived the blast would be re-boarded and sailed home. A menagerie of experimental animals – 200 goats, 200 pigs, and 5,000 rats – locked up on the ships of the target fleet would yield data on the biological effects of radiation.

The military planners were also interested in the psychological effects of the explosions. According to the official history of the tests, four of the goats were chosen 'because of their psycho-neurotic tendencies'. Troops also would be required to witness the tests, and their reactions would help planners predict how soldiers would behave in a real atomic war.

Crossroads' second purpose was to win friends for atomic testing. Unlike its wartime predecessors, Crossroads was anything but a secret operation. More than 150 journalists, broadcasters, and photographers were flown out to Bikini to report on the tests, which were broadcast live from the USS *Appalachian*. Sixteen senators and representatives, and a sizable contingent of non-congressional bigwigs, went to Bikini on what *Newsweek* termed a record-breaking junket. Truman, in fact, ordered the tests postponed for six weeks so that the mass exodus from the capital would not endanger his legislative agenda. The President, noted his Secretary of Commerce, Henry Wallace, 'didn't want a lot of Democratic congressmen out witnessing the test when their votes were needed here in Washington'.

Several months before Crossroads was to begin, Stafford Warren assembled a 60-man radiological safety team. Half were experienced Manhattan Project health workers. The rest were military doctors who received an intensive two-and-a-half-month tutorial in physics and radiation safety. The Radiological Safety Section – RadSafe for short – had the dual job of ensuring the safety of the test personnel, and enabling them to return to the target fleet as soon as possible after the test. Its procedures and standards were reviewed by the Medico-Legal Board, a military group established to 'reassure Col. Warren that . . . no successful suits could be brought on account of the radiological hazards of Operation Crossroads'.

At almost the last minute it became clear that Operation Crossroads had dramatically expanded. Warren and his staff suddenly realized that they had not enough men and not enough equipment to ensure the safety of the tens of thousands of troops and civilians who would now be taking part. As Warren wrote later, 'the test turned out to be literally a hundred times larger than the original conception'. He hurriedly recruited all available Manhattan Project health workers, including those who had already returned to civilian life. Even this was insufficient, however, and several hundred totally inexperienced men were recruited to RadSafe. The only training they received was a series of shipboard lectures en route to Bikini.

For months before the operation, national magazines and newspapers carried stories emphasizing the contributions Crossroads would make to science and medicine. Though the Joint Chiefs of Staff had made it clear to Admiral Blandy, the Crossroads commander, that military aims would take precedence over science, *Newsweek*, in a story called 'The Good That May Come From The Tests At Bikini', informed its readers that the operation was not intended as a show of military force, but as a scientific experiment. 'The tests on animals, at varying distances from the explosion, should be especially valuable, through their contribution to medical knowledge.' *Science News Letter*, too, stressed the humanitarian aspect of the tests, which would show 'whether more sensitive or more exact devices may be needed to indicate quickly enough the need for special medical treatment of atom bomb victims'.

A secondary theme of the Crossroads publicity was the possibility that the tests might trigger a cataclysmic tidal wave or earthquake. Never before had an atomic bomb exploded underwater. There were predictions that the blasts would cause the Pacific Ocean to drain through newly created fissures into the earth's crust; that they would trigger cataclysmic tidal waves, destroying the entire fleet and going on to drown coastal cities from Vancouver to New Zealand. The authorities objected to these dire predictions. Blandy vehemently denied charges that the tests were irresponsible, saying, 'I am not an atomic playboy.' The widespread anticipation of disaster was to work in his favour, for the first test, Able, was an anti-climax.

At 9.00 on the morning of 1 July 1946, bomb Able, adorned with a picture of Rita Hayworth, was dropped from a B-29 flying 30,000 feet above Bikini lagoon. A brilliant light flashed across the sky. At the same time, a churning mass of glowing gas burst into a huge ball of fire. The path of the shock wave moving out from the centre of the blast was visible on the water's surface. A giant mushroom cloud emerged from the ball of fire and rose tens of thousands of feet above the lagoon. It was spectacular, but for those expecting the end of the world, or at least an all-engulfing tidal wave, it was a let-down. David Bradley, an Army doctor recruited to Crossroads' radiological safety section, had a bird's-eye view of the test and its aftermath as he waited to fly through the mushroom cloud in a radiological reconnaissance plane. The lagoon was calm. A few ships were on fire, a few showed visible damage, but on the whole the fleet had survived the blast well. 'Expecting much more dire and dramatic events, our crew was disappointed,' Bradley wrote. 'There was much pooh-poohing of the Bomb over the interphone, and at last the co-pilot growled: "Well, it looks to me like the atom bomb is just like the Army Air Force – highly overrated."'

The radio audience back home was equally disappointed. Able didn't make as much noise, or cause as much damage, as people expected. A telephone survey conducted in Philadelphia found that the Crossroads radio show lost many of its listeners to the broadcast of a baseball doubleheader. The news stories filed from the USS *Appalachian* had the overall effect of reassuring people that atomic bombs were perhaps not as bad as they had believed. 'Awful as it was, it was less than the expectations of many onlookers,' was *Time*'s comment. Returning to the same theme the following week, the news magazine philosophized, 'The Thing had grown a little less awful as a result of Bikini, its apparently infinite power was finite after all.' The *New York Times*'s article was headlined 'Bikini "Dud" Decried For Lifting Fears'. The reporter, William Laurence, who had witnessed both the Trinity and the Nagasaki explosions, reflected upon the public's eagerness to belittle the bomb. 'Having lived with a nightmare for nearly a year, the average citizen is now only too glad to grasp at the flimsiest means that would enable him to regain his peace of mind.'

Able pleased the test authorities. Neither blast nor fallout were as bad as feared. RadSafe was especially relieved that Able had not lived up to some of the lurid predictions made about it. One of Blandy's experts, Captain A. A. Cumberledge, had warned that Able would produce 'the equivalent of tons of radium floating loose in the atmosphere in deadly concentrations', which, unless carried away by the wind, would seriously contaminate the fleet. But the weather was kind on Able day: winds dispersed the bomb's radioactive cloud high in the atmosphere more quickly than expected. At Rongerik, 100 miles to the east, the transplanted Bikinians, who had been held on ships during the blast in case a sudden evacuation proved necessary, were returned to their new home only an hour after the explosion.

Within four hours of the detonation, RadSafe teams had declared several target ships radiologically safe – 'Geiger sweet' in test parlance – and allowed fire-fighting and salvage teams to board them. By 2.30, less than seven hours after the test, Blandy, on the advice of RadSafe, announced that the lagoon was safe for all ships. The live fleet, which had moved out into the ocean for the test, returned to the lagoon. By the end of the second day, only ten of the target ships were still unfit to be boarded, due to radioactivity or structural damage.

The results of personnel monitoring were equally reassuring. All units had been at least 15 miles away from the blast, a fact that caused some troops to complain that they had been deprived of a good view of the blast. Fewer than ten per cent of the 42,000 member task force had been issued film badges, but only ten badges showed readings at or above the limit for gamma radiation, which was set at 0.1 r per day, as it had been during the Manhattan Project. No badge recorded more than 0.2 r per day. RadSafe's film badge expert announced that 'no personnel of the task force obtained a physiologically significant radiation dose'. Apart from the fact that the bomb had been dropped almost two miles off target, Able had been, in the words of the official testing history, 'a perfect test'. The task force turned to Baker, and the world turned back to business.

Thursday 25 July was Baker Day. David Bradley, again in a reconnaissance plane, witnessed the blast from the air 15 miles to one side of the target. The explosion 'seemed to spring from all parts of the target fleet at once. A gigantic flash – then it was gone. And where it had been now stood a white chimney of water reaching up and up. Then a huge hemispheric mushroom of vapour appeared like a parachute suddenly opening. It rapidly filled out in all directions. By this time the great geyser had climbed to several thousand feet. It stood there as if solidifying for many seconds, its head enshrouded in a tumult of steam. Then slowly the pillar began to fall and break up.' As the pillar collapsed, dropping radioactive rain into the lagoon, a dense wall of steam, a thousand feet thick and three miles in diameter, rose from its base. It surged through the lagoon, smothering the target fleet and moving towards the surrounding islands.

Less than an hour later, Bradley's and another reconnaissance plane flew over the lagoon, taking readings. They encountered such high levels of radioactivity that they were ordered to turn back. 'We were thankful for that,' Bradley wrote. 'It is not that we were in any immediate danger. But with radiation so intense at such an altitude, that at water level would certainly be lethal. And this wasn't just a point source; it was spread out over an area miles square.'

Two hours after the explosion, several patrol boats carrying RadSafe monitors entered the lagoon and approached the target fleet. The radiation readings at the outskirts of the fleet were so high that the boats could go no further. One was so severely contaminated by the brief foray that it had to be temporarily abandoned. By nightfall, almost the entire lagoon was still too hot to enter. Blandy informed the Joint Chiefs of Staff in Washington that 'radioactivity persisting in water and on board' might delay boarding of the target ships for several days. Not only was the water radioactive; the tons of coral pulverized and thrown into the air and on to the ships by the explosion gave off very high readings. It soon became clear that virtually the entire target fleet was severely contaminated.

The test planners were taken aback. They had not expected such high levels of radioactivity, though there had been warnings from RadSafe. Both Able and Baker had added huge amounts of radioactive fission products, the equivalent of tons of radium, to the environment. But Able's radioactivity was released to the open air and dispersed over a wide area. Baker's was trapped in the tiny lagoon, creating a concentrated pool of poison that contaminated everything it sprayed over or lapped against, including the US fleet. One pre-test report had warned that 'On some target ships, possibly within 1,000 yards of detonation, boarding inspection may be dangerous for weeks.' The lagoon was so radioactive that for three days after the test, most of the live fleet was obliged to wait idly at the entrance, watching the damaged ships fill with water and sink. Only a few vessels were allowed in, to gingerly take readings before rushing back out. By the fourth day, radiation levels were down somewhat, and Blandy ordered the live fleet into the lagoon for a massive salvage operation. The salvage teams wore film badges and were accompanied by RadSafe monitors, whose job it was to ensure that the workers did not exceed the daily limit for gamma radiation of 0.1 r. In many cases it took only 20 minutes to reach the daily limit.

The task force now faced a huge and unexpected job – decontaminating the target fleet so it could be re-boarded and sailed home. Unfortunately no one knew how to clear a ship of radioactivity. In the first days, crews simply sluiced down the decks of their ships, using radioactive lagoon water. When that didn't work, they used soap and water. That too failed, as did every other cleaning agent tried, from lye to foamite. In fact, experiments indicated that all the washing and scrubbing had only succeeded in soaking the fission products more deeply into the ships' wooden decks. After many weeks it was finally proved that the only effective decontamination technique was to remove the entire outer surface of each ship to a depth of almost half an inch. Sandblasting was tried, but, as Bradley pointed out, it was hardly practical in an isolated tropical lagoon. 'It is tedious, hot work, and since there may be a danger from inhaled radioactive dust, it has to be carried out by men fully clothed and working in oxygen re-breathing apparatus. Next to a diver's suit, there is nothing more cumbersome and trying on the nerves than such a costume in the tropics.' Strong acid or planing might also work on a small scale, but there was no way to scrape down the entire fleet.

Many of the inexperienced monitors lacked the confidence to stand up to officers whose overriding concern was to board the contaminated vessels as soon as possible and who lacked a healthy respect for the invisible dangers of radiation. According to David Bradley, a large proportion of the 400-strong RadSafe unit 'did not have the authority to order a whole bunch of sailors off a boat'. RadSafe monitors complained that many of the officers in charge of the working parties had a 'hairy-chested' attitude toward radiation. 'It appears,' reported the monitors assigned to the German battlecruiser *Prinz Eugen*, 'that there is an attitude of indifference on the part of the ship's officers . . . to the safety standard set by RadSafe. There is reason to believe that men are being kept aboard for longer periods than they should be and also that [the officers assume that] the standard of 0.1 r has such a large safety factor that it can be ignored.'

The task force was not prepared for large-scale decontamination work. Sailors were not issued proper protective clothing – a garment to cover the entire body and head, along with goggles, boots, gloves, and filter masks – while working on contaminated vessels. The first clear order to destroy severely contaminated clothing was not issued until two weeks after Baker was exploded. Not until 13 August, almost three weeks after the blast, were decontamination crews ordered to board a 'change ship' to shower and change their workclothes before returning to the ships where they slept and ate. But by then Blandy had already realized that Operation Crossroads would have to be called off. It was simply too dangerous.

Blandy's realization began on 7 August, when he received a chilling private message from Stafford Warren. 'The target vessels are in the main extensively contaminated with dangerous amounts of radio-activity,' Warren wrote. 'Quick decontamination without exposing personnel seriously to radiation is not possible under the present circumstances and with present knowledge.' Warren continued, 'the

95

present operations in the Bikini lagoon should be terminated on or by 15 August 1946, since there is neither equipment or monitoring personnel available to continue safety operations beyond this date'. To his wife he wrote, 'I'm getting the jitters, & so are the rest because there are so many men here, altho they are leaving in large no's. daily.'

On 10 August, Admiral Blandy abandoned Operation Crossroads. He sailed for Pearl Harbor, leaving behind only a skeleton task force – including RadSafe – with orders to do the minimum decontamination work necessary to enable the target fleet to be towed away from Bikini. Bradley reflected upon this turn of events. 'The target fleet which was to have steamed triumphantly back to Pearl and the Golden Gate, invincible as ever, will remain here at anchor, blackened with flame and streaked with foamite and rust, until the ships can be safely disposed of. Some may eventually be towed to the West Coast; many will certainly be sunk in deep water.'

The skeleton task force stayed at Bikini until the end of August, trying to decontaminate the surviving target ships. In the end, however, only eleven ships of the target fleet were able to leave Bikini lagoon under their own power – and five of them hadn't been contaminated in the first place. The rest, about 50 vessels, had to be towed out, because they were too radioactive to be manned. It was not deemed safe for the target ships to return directly to the United States, so they were taken to Kwajalein, the largest of the Marshall Island atolls. Forty-two of the target vessels were sunk there. Twelve target ships were towed back to shipyards in Puget Sound and San Francisco, for decontamination experiments and for use in training radiation monitors.

The target ships were not the only ones made radioactive by Baker. The live fleet, the ships on which most of the 42,000 participants were sleeping, showering, and eating, had also become contaminated, largely as a result of entering the lagoon prematurely. On the day the live fleet re-entered the lagoon, Bradley found 'that the [lagoon] water was steadily increasing in radioactivity ... by noon the intensity was such as to endanger our water intakes and evaporators, and so at the request of the Radiological Safety Section the entire live fleet up-anchored and sailed to a point nearer [the entrance to the lagoon]'. But the water was radioactive throughout the lagoon. It splashed over the ships, even contaminating the algae that covered the ships' hulls. Warren recommended that all drinking water should be taken from the ocean, as far as possible away from the lagoon, but his advice was ignored. The radioactive lagoon water contaminated the evaporators used to collect it and the pipes that carried it to the showers and toilets. As long as the task force remained at Bikini, the contamination worsened.

One of the biggest problems was that RadSafe had no instruments that could detect alpha radiation in the field. The RadSafe laboratory at Kwajalein had only one alpha counter, according to which there was, as Bradley noted, 'a small but definite amount of plutonium spread atom-thin over most of the contaminated areas'. For the most part monitors could only guess about the presence of plutonium. There *were* instruments capable of detecting gamma and beta rays, but even these were unreliable and prone to breaking down in the humid weather and rough field-conditions of Bikini.

In addition, President Truman's insistence that the tests be post-poned for six weeks meant that more than 300 experienced monitors, recruited from academia, had to go back to teach their fall courses during the crucial post-Baker decontamination period. Attempts to get these men to stay beyond the periods of their contracts, 'were met with unanimous refusal', reported Warren. The several dozen experienced monitors who remained were assisted by troops given on-the-spot training.

Though only a small percentage of Crossroads participants were issued film badges, which gauge gamma exposure, the readings were reassuringly low. From the day of the Baker explosion until the task force left Bikini in late August, 6,000 badges were issued. Most read zero. Fewer than ten per cent recorded exposures greater than the daily limit of 0.1 r. A 1985 study by the US Government Accounting Office found, however, that the Crossroads film badges could not record exposures of more than 2 r, and exposures of half the daily limit or less appeared as zero. On Baker Day, when Bradley's reconnaissance airplane was ordered down because of the dangerously high levels of radioactivity in the cloud he was sampling, he wore two film badges, neither of which detected any radiation. Using the monitoring equipment on the plane, Bradley was able to compute the dose he received that day at somewhere between 2.7 and 17 r.

Inexperienced safety personnel, inadequate equipment, and difficult working conditions, obscured the emerging picture of a seriously contaminated live fleet. Warren's private papers, which became avail-able after his death in 1982, reveal the severity of the situation. On 13

August, he reported that 'The initial contamination of surfaces was so great that reduction ... of 90 per cent or more still leaves large and dangerous quantities of fission [products] and alpha emitters scattered about ... Contamination of personnel, clothing, hands, and even food can be demonstrated readily in every ship ... in increasing amounts day by day.' Even Warren's headquarters, the USS *Haven*, was not safe. Bradley noted that 'the evaporators aboard the USS *Haven* have become quite hot in themselves, the scale inside the evap[oration] tanks acting as a sponge for radioactive particles. Similarly the salt-water lines of the ship (fire lines, toilets, and so on) show the presence of radioactivity, too, and must be checked every day or two ... the waterlines remain so hot that one may pick up a respectable radiation coming through the steel walls of the hull.' Some of the hulls were so radioactive that the bunks in which the sailors slept had to be moved away from them.

In an interview 40 years after Crossroads, Bradley recalled the impossibility of avoiding radioactivity in everyday activities. 'I drank milk instead of water, when I could get it. But you had to drink water in that hellish hot place ... Our real worry was in breathing, and you have to breathe ... I don't think we understood the degree to which this stuff was spread all over, from the decks of the ships to the corridors of the ships, the gangways, the mess halls, the bunkrooms, and the toilets. I didn't realize it until about four years ago, when I read Stafford Warren's papers.'

By the time the true extent of the live fleet's contamination was acknowledged, the fleet had already dispersed. Some ships had sailed for Pearl Harbor with Blandy, some were heading for mainland home ports; some were still at Bikini and Kwajalein. Belatedly, orders went out to check all vessels. In September 1946 the Navy decided every ship that had been at Bikini during or after the Baker test had to undergo full-scale decontamination. Some ships were returned to active service after being sandblasted or otherwise decontaminated at Navy ports around the country. Others were put into mothballs, where they still remain. Still others were sunk. But the Navy could not afford to have an entire fleet out of action for months while a way to clean it was sought. 'Consequently,' an official report stated in September 1946, 'several APA's, Destroyer Division 72 and some auxiliaries have been cleared practically to meet operational requirements on the basis that they might as well continue to operate until methods of making them safe for overhaul are developed.'

While the operation continued, sailors ate, worked, and slept on the contaminated ships. New restrictions were imposed on the seamen. 'To the sailors it must all be incomprehensible,' reflected Bradley. 'Here we are insisting on measures of extreme caution at a time when the radiation has had a month and a half to decay away, when the ship's decks are probably not a thousandth as hot as they were in the first hour following the Baker exploision. These same measures were not employed in the first weeks.' A number of ships, including the USS *Clymer*, were stopped mid-route and ordered to throw all food supplies overboard. Troops on the *Clymer*, worried that they might have eaten contaminated food, say that they were warned by their superiors not to mention the incident.

Officially there were no radiation casualties at Crossroads, but a number of seamen believed that they had been injured, and rumours of radiation injuries circulated through the fleet. On the day after the Baker test, sailor Frank Karasti was ordered to prevent the destroyer *Hughes* from sinking. 'Out of the four hours we spent on her, two were spent vomiting and retching as we all became violently ill,' he told writers Harvey Wasserman and Norman Solomon. One month later lesions appeared on Karasti's lungs, and he has since been afflicted with a number of other health problems – deteriorating skin, difficult breathing, and hypertension – that he attributes to his Crossroads days. Shortly after Baker, seaman Richard Stempel was 'swimming nearly every day [in the target area] and using the water freely. We were never told not to do either.' One time Stempel and three friends were sitting on a buoy moored in the target area, when an officer with a Geiger counter came by. 'The Geiger counter pegged and he ordered us off. He didn't advise us any decontamination procedure.' A few weeks later Stempel developed a skin disease that the ship's doctor could not diagnose. Forty years after Crossroads tests, John Grifalconi, assigned to the operation as a photographer, went to a reunion of sailors and other people involved in the tests. 'So many of them were ill,' he later recalled, 'that I almost felt guilty that I had nothing worse than cataracts. However, last Christmas I discovered that I hadn't escaped. I was diagnosed as having a very serious form of leukemia.' Grifalconi is tall and very thin; his duck-billed cap covers a bald head; his skin is the pale yellow of the chemotherapy patient. Neither his illness, nor those of his former colleagues, are considered by the US government to be linked to their time at Bikini.

The top-secret post-Crossroads Decontamination Report reported that General Groves 'is very much afraid of claims being instituted by men who participated in the Bikini tests'. Publicly, however, there were no such fears. Admiral Blandy announced with pride that Crossroads caused 'no casualties from overexposure to radiation . . . In fact, there is no evidence that any person suffered any ill effects whatever from either of the two atomic bomb explosions at Bikini.' The nation was happy to believe him. As the historian Robert Jungk wrote 12 years later, the 'spiritual effect [of the Crossroads tests] was great. For they soothed the fears of the American public almost as much as the bombs dropped on Japan had aroused them.'

CHAPTER ELEVEN

The Proving Ground

One year after the Crossroads tests, a rash of anniversary articles caused a revival of atomic anxiety – with a new twist. For the first time, the focus of fear was radiation rather than the bomb itself. In language that the public could understand, the reports said that there is no protection against the insidious spread of the radioactive poisons created by atomic explosions. Especially influential was a piece in *Life* magazine by Stafford Warren, the man responsible for radiation safety during Operation Crossroads. Warren wrote frankly about the extent and seriousness of the radioactive damage caused by the Crossroads tests. He reported that the test fleet 'will remain contaminated for years'; that '300 men . . . lived and worked in the heavily contaminated area'; and that even the active fleet became so contaminated that 'in some instances it became necessary to move bunks back from the hulls to protect the men who slept in them'. Readers were appalled to learn that radiation at Bikini was so pervasive that RadSafe workers occasionally 'had to dissolve the outer layer of [a worker's] skin from his hands with acid'. Nonetheless, Warren assured his readers, 'Not one of the 42,000 men who went to Bikini had been detectably injured by radiation.' Despite his frightening revelations, Warren's message was a reassuring one. Radiation from atomic weapons is deadly, but – in peacetime, at least – the experts can keep it under control.

In the fall of 1948 David Bradley published *No Place to Hide*, his account of the Crossroads tests. Written as a diary, the book is a straightforward record of what he saw and did during his tour of duty. Bradley writes movingly of the beauty and peace of the South Pacific islands where he lived and worked for four months. His horror as he gradually realizes the extent of the radioactive contamination that the tests have caused is the more powerful because he expresses it as an affectionate observer not as a scientist or a polemicist.

Bradley's chief message – there is no defence against radiation –

gave a boost to the cause of placing atomic weapons under international control. 'What happened at Crossroads,' Bradley wrote, 'cannot be buried with the ships in Bikini lagoon or towed away to rot on the beaches of Kwajalein. What happened at Crossroads was the clearest measure yet of the menace of atomic energy. Less spectacular perhaps than Hiroshima and Nagasaki, the Bikini tests give a far clearer warning of the lingering and insidious nature of the radioactive agent.' *No Place to Hide* was an immediate critical and commercial success. It was on the *New York Times* best-seller list for ten weeks. One-quarter of a million copies were sold in its first year. It fuelled alarm about the bomb, and specifically about atomic radiation.

The US military and others involved with the testing programme were alarmed by the public impact of Bradley's book. Government agencies and individuals with links to the testing programme mobilized to counter public fears about radiation. The counter-attacks had three main themes: 1, radiation is natural; 2, the public is protected from it by high safety standards administered by well-trained scientists; and 3, fear of radiation is more dangerous than radiation itself.

In a not entirely successful attempt to defuse concern, the US Surgeon-General in 1948 released a statement entitled 'Army Doctors Say Hysteria Need Not Follow Atom-Bomb Explosion'. Decrying 'sensational prophecies' that radiation would cause birth defects, the report said that because radiation is usually fatal to developing embryos, there would simply 'be a higher rate of abortion and miscarriage'. Neither should hair loss caused by radiation be a worry, the doctors counselled. It will grow back eventually, 'if the patient has not received a lethal dose of radiation'. *Time* magazine headlined its article about the report, 'Feel Better Now?'

Shortly after Bradley's book appeared, Colonel James Cooney, an Army doctor who had been at Crossroads, gave an address to the American Public Health Association in which he argued strongly that soldiers had to accept radiation exposure as just another hazard of battle, or they would not be able to fight the next war. As things stood, Cooney said, 'the fear reaction of the uninitiated is appalling'. Cooney's talk illustrated a significant change in the public's attitude toward the bomb in the three years since its debut. Cooney deplored the fact that many people, civilians and military men alike, had forgotten that 'the atomic bomb was developed as a blast weapon'. Instead they had come to 'believe that the bomb is primarily a weapon which destroys by mysterious radioactivity'.

The following year, Ralph Lapp, a physicist with the Office of Naval Research, the Navy's scientific think-tank, who had also served at Crossroads, published *Must We Hide?*, a riposte to *No Place to Hide*. Lapp, whose book, like Bradley's, was directed at the lay reader, argued that the hazards of radiation are no different than those of any other modern discovery, and should not be exaggerated. Radiation, like sunlight, can cause damage, but 'we are fortunate in having a nucleus of well-trained scientists who have a thorough understanding of radiation hazards and methods of protection against them'.

The onslaught on Bradley's book – and on the thinking behind it – continued. In 1950 the first nuclear-age civil defence manual appeared. *How to Survive an Atomic Bomb*, by Richard Gerstell, another Crossroads alumnus, was written at the suggestion of Secretary of Defense, James Forrestal. Using a question-and-answer format, Gerstell informed readers that 'you can easily protect yourself' from radiation, and that 'much of the danger from radioactivity is mental'. *How to Survive* was the first popular publication to use the term 'fallout', though the author expressed his disdain for the importance of the concept by referring to 'that "fallout" stuff'. 'Radiological defense, which consists of the detection and avoidance of radioactive hazards, is something which, in a quiet but effective way, the government has been perfecting for years,' he assured readers. Gerstell concluded that radiation is 'much the same as sunlight'; that in ordinary doses it carries no danger; and that the experts have things under control.

By 1950 the public had largely accepted the government's line on radiation. The prevalent attitude was that it was unreasonable, dangerous, and perhaps traitorous, to rail against radiation and the bomb.

The early years of atomic testing were quiet ones. In the first five years after the war, the Americans exploded only five bombs, all in the South Pacific – and the Russians only one, in Siberia in August 1949. The outbreak of war in Korea disrupted US plans for another Pacific test series early in 1951. For a combination of logistical and security reasons, the US government decided it needed a domestic testing ground. After briefly considering a site on the coast of North Carolina near Cape Hatteras, a selection committee settled upon a 1,350-square-mile government-owned tract of the Nevada desert. The Nevada Proving Ground, or Nevada Test Site as it has come to be called, is situated 65 miles northwest of Las Vegas, then a town of 25,000 inhabitants. Prevailing winds blow northeastward, away from Los Angeles and Las Vegas, and towards rural areas of Nevada, Utah, and northern Arizona, where about 100,000 people lived.

On 1 August 1950 Edward Teller, Enrico Fermi, Alvin Graves, the director of the AEC's test programme, and other atomic experts held a secret daylong meeting at Los Alamos to discuss the radiation problems that exploding atomic bombs in Nevada would cause. A meeting of radiation protection specialists from the United States, Britain, and Canada had the previous year recommended that the public not be allowed to receive more than three millirems (0.003 rem) of external radiation per week. But the AEC experts decided that members of the public would not be evacuated unless they were exposed to more than 25 rem in four weeks. This was 20 times more radiation than AEC workers were allowed to receive over a four-week period, but officials reasoned that workers were continually exposed to radiation and members of the public were not. It was 'generally accepted by the medical profession that this amount of radiation will cause no disability, regardless of the physical condition of the recipient,' said James Cooney. The committee did agree that 25 rem is 'perhaps a little more radiation than medical authorities say is absolutely safe – six roentgens'. Two years earlier the NCRP had concluded that there is no 'absolutely safe' level of radiation exposure. The scientists concluded that, given favourable weather conditions, bombs up to 25 kilotons could be exploded without harm to the public, but less than four months later the AEC commissioners decided to test weapons of up to 50 kilotons in size at the site.

Just before Christmas 1950, military and AEC staffers met to consider how best to introduce the site to the public. The central issue, according to minutes of the meeting, was how 'to make the atom routine in the continental United States and make the public feel at home with atomic blasts and radiation hazards . . . It appeared that the idea of making the public feel at home with neutrons trotting around is the most important angle to get across.' The AEC worked hard to get that message across, sending teams of public relations men to communities near the test site, giving public talks, distributing AEC booklets, and showing AEC films.

Because Las Vegas relied heavily upon tourism, the city was eager to cooperate in dispelling worries about the tests. The Chamber of Commerce circulated humorous leaflets about the testing, one of which featured a Bikini-clad glamour girl holding a Geiger counter to an old uranium prospector. Local businesses, entering into the swing of things, invented an Atomic Hairdo and an Atomic Cocktail made of equal parts of vodka, brandy, and champagne, with a dash of sherry. 'The angle,' an official told writer Daniel Lang in the early 1950s, 'was to get people to think the explosions wouldn't be anything more than a gag.' Given the unusual nature of Las Vegas's tourist industry, the bomb actually proved to be good for business. As one hotel manager remarked, 'Before the proving ground, people just heard this was a wide-open town. Now that we're next door to the atom bomb, they really believe it.'

Most of the 100,000 people who lived downwind of the test site were Mormons, rural or small-town folks whose lives were centred on the church. Life in these communities, observes Howard Ball, professor of political science at the University of Utah, 'was Church, Flag, Mother, Apple Pie, and Chevrolet: a religious, patriotic, ordered, and simple one'. Local people certainly didn't like atomic bombs any more than anyone else, but they accepted the government's assurances that the tests were controlled, scientific experiments, conducted under strict conditions of safety. Most downwinders were proud that the atomic testing programme was located in their territory, that they were involved in something so important to the defence of their country and the defeat of Communism, and they were willing to do their share to make the programme a success.

At dawn on Saturday 27 January 1951, a brilliant flash of light lit up the Nevada sky, as the first atomic bomb exploded at the test site. Many Las Vegas residents were awakened by the shock wave that followed seven minutes later. A second bomb was detonated the next day, and three more over the next eight days. Several of the explosions lit up the night sky over San Francisco and Los Angeles, hundreds of miles away. The only casualties of the series were some windows, mirrors, and walls in Las Vegas that were cracked by the explosions' powerful shock waves.

As the tests continued, the trepidation of nearby residents settled into a sort of proprietorial nonchalance. Nevada's governor, Charles Russell, had his worries about the possible ill effects of the tests, but was nonetheless proud of Nevada's contribution to the nation's security. 'No matter who's right, it's exciting to think that the submarginal land of the proving ground is furthering science and helping national defense. We had long ago written off that terrain as wasteland and today it's blooming with atoms.'

The AEC promoted bomb-watching as an entertaining and educational holiday activity. An article in the travel section of the *New York Times* described such an outing. 'In the wake of a detonation, the atomic cloud can be seen attenuating across the sky. It may come over an observer's head. There is virtually no danger from radioactive fallout.' For local families, school groups, and tourists, a pre-dawn expedition to Mount Charleston, 50 miles from the test site, became a popular outing.

From a public relations point of view, the first two test series, Ranger and Buster-Jangle, went off well. (Test names were assigned from a government list of words suitable for use as meaningless code-names for sensitive projects.) But the third test series, Tumbler-Snapper, conducted from early April to early June 1952, caused some anxiety. Unexpectedly strong winds on 7 May, the day of Easy, Tumbler-Snapper's fourth shot, 'strewed heavy fallout well beyond expected ranges', according to the official history of the testing programme. A secret AEC report found that Easy had caused 'the largest fallout we have ever had over a populated area' since Trinity.

The fallout only came to light when a radio technician named Lyle Jepson won $10 in a competition for news tips sponsored by the Mormon Church's newspaper, the *Deseret News*, by reporting high levels of radioactivity in Salt Lake City, 400 miles northeast of the test site. The AEC acknowledged the high readings, but said that they posed no danger. The commission issued a press release stating that 'fallout readings . . . were primarily of scientific interest'. In an editorial on the affair, the *News*, which like virtually every other local paper strongly supported the testing programme, accepted the commission's assurances, but predicted that 'Salt Lakers . . . will now be more keenly aware of potential deadliness in the air each time a nuclear device is exploded', and warned the AEC that 'precautions must be redoubled, never relaxed'. It was the first glimmer of doubt about the commission's role as guardian of the public health.

Some Las Vegans were also put out by Tumbler-Snapper, but for a different reason. Several people complained to the local AEC office that the eight shots had not been dramatic enough, and a columnist for the *Las Vegas Review-Journal* accused the commission of 'gypping its public' with Tumbler-Snapper's measly display. 'Bigger bombs, that's what we're waiting for,' said one Las Vegas nightclub owner. 'Americans have to have their kicks.'

In late 1951 a new element was introduced into the Nevada test programme. The Army set up a bivouac, Camp Desert Rock, next to the test site, so that thousands of troops, from all branches of the services, could witness the tests and conduct field exercises during them. Record-keeping was poor, but the Defense Department now estimates that during the period of atmospheric testing, from 1945 to 1963, between 250,000 and 500,000 troops took part in tests of atomic weapons. 'Indoctrination in essential physical protective measures . . . and observation of the psychological effects of an atomic explosion are reasons for this desired participation,' the Pentagon's Military Liaison Committee informed the AEC. After witnessing the tests or taking part in war games, Desert Rock troops were given psychological tests. To gauge the physical effects of the tests, the Army also stationed pigs, rabbits, and sheep, some in custom-made uniforms, at various distances from ground zero. On one occasion a group of pigs outgrew their uniforms as a result of a delay in the testing schedule and had to be outfitted in new ones.

The Army wanted the training to be as realistic as possible, and that meant the troops experiencing fallout. At an AEC meeting in November, 1957, Libby told his fellow commissioners that the Army had reservations about underground testing and fallout-free bombs, which the AEC then thought it could develop. Libby said he 'understood there was a desire by the military for some degree of off-site radiation for troop training purposes'.

The exposure limit for troops was initially set at one rem for an entire test series, but in mid-1952 it was raised to three rems per test series, and still later to six rems per series. Troops were generally not issued protective clothing or equipment such as gas masks or film badges. Officials claim that every military participant was issued a film badge, but testimony by troops indicates that from one-third to two-thirds did not receive film badges. The Army provided its own RadSafe group, consisting of troops from units as diverse as the 216th Chemical Service Company and the 995th Quartermaster Laundry Company. Few of the monitors had ever witnessed an atomic explosion; even fewer had field experience in radiation protection, though they did receive several weeks of training.

At first the AEC kept the troops at least seven miles away from ground zero during shots, but in 1952 the military threatened to boycott the tests if troops weren't allowed closer in. An AEC memorandum records that on 25 March 1952 'the commission then discussed the necessity for realistic training by the military in all fields, often accompanied by serious injuries, and that such training was also necessary in the field of atomic weapons'. Air Force Brigadier General A. R. Leudecke told the AEC that 'the tactically unrealistic distance of seven miles' was making soldiers nervous about the bomb and causing 'unfavorable psychological effects'. Shields Warren, director of the commission's Division of Biology and Medicine, was against allowing the troops any closer to ground zero. He warned that there would be 'larger numbers and more serious casualties the closer the troops were to the point of detonation'. Nevertheless, after two days of discussion, the commissioners agreed to let soldiers be stationed 7,000 yards (less than four miles) from ground zero during tests. The commissioners refused, however, to take any responsibility for troop safety at that distance and demanded that the Department of Defense 'be informed of the possibility that exceeding the normal limits of exposure to radiation or pressure might endanger the participating personnel', and 'that [the military's] acceptance of responsibility be adequately documented'.

Many of these so-called atomic veterans now have health problems which they believe were caused by their exposures to radiation while in the service. Six thousand of them have filed claims with the Veterans Administration, but all but 44 have been denied on the grounds that there is no evidence linking radiation exposure to health damages. In 1988 the US Congress passed a bill requiring the government to grant health benefits to any atomic veteran who has contracted a type of cancer that is known to be caused by radiation.

Upshot-Knothole, the inelegantly-named fourth test series, was a watershed for the AEC, and for nuclear testing worldwide. At the outset of the series, in March 1953, the atomic testing programme enjoyed widespread public support. Three months later, when the problem-ridden series was finally over, the public had learned to fear fallout and the future of the test programme was in jeopardy.

The first sign of trouble came on 25 March 1953, eight days into Operation Upshot-Knothole, when Dr Lyle Borst, chairman of the

University of Utah's physics department and previously a director of the AEC's Brookhaven National Laboratory, publicly expressed concern about fallout from the tests. Borst, who lived in Salt Lake City, told the *Deseret News* that he did not allow his children to play outdoors after an atomic test and that he made them wash more often during test periods. 'I would no more let my children be exposed to small amounts of radiation unnecessarily than I would let them take small doses of arsenic,' he told the *News* reporter. The paper's editorial page took Borst's fears seriously: though the AEC has 'given assurance that the level of radiation experienced so far cannot possibly be harmful to human beings', it stated, there is cause for concern, especially for residents of towns that 'fall within the 200-mile circle around the proving grounds which is covered by a security blackout. The public is never told just what levels of radiation are reached in this area.'

Borst had touched a sensitive nerve in the family-oriented Mormon farming communities. Nothing could alarm them more than the thought that fallout might be harming their children or that their own exposure to radiation might result in a legacy of disease and mutations for their descendants. As writer Daniel Lang remarked in a 1955 article about fallout for the *New Yorker*, 'in these times, as the AEC knows all too well, the subject of mutations is a touchy one. Many people find it much easier to contemplate – in theory at least – the possible destruction of the world while they themselves are still inhabiting it than to reflect that their descendants, centuries hence, may inherit genes that were impaired by the current tests.'

It had been known for many years that radiation can damage reproductive cells and cause genetic mutations in the offspring of exposed individuals. Hermann Muller of Indiana University was awarded the 1946 Nobel Prize in medicine for his discovery in the 1920s that radiation caused genetic mutations in fruitflies. In 1946 *Time* magazine took a light-hearted view of mutations, predicting that one result of increased exposure to radiation would be that 'more redheaded children will be born in black-haired families, and more mutations will lurk in the germ plasm to scandalize future neighbors . . . Some geneticists believed that the next Hiroshima generation or so would show many bad mutations. But most would be eliminated by the law of the survival of the fittest. The few superior mutations would survive to improve the race.' But, in an interview after accepting his Nobel Prize, Muller himself told *Time* that mutations were much more likely to be negative than positive. 'Good [mutations] are so rare that we can consider them all as bad,' he said.

Exactly one month after Borst's interview, the AEC detonated a bomb bigger than any previously exploded on American soil. At 43 kilotons, test Simon was almost double the 25-kiloton limit set for the Nevada site by the AEC's own advisers. But it was the weather as much as Simon's size that caused the trouble that followed. According to a now-declassified secret test report, 'the amount of fallout was expected to be much larger than usual. However, due to the fact that no populated communities were expected to be in its path, the decision was made to fire on schedule.' It was AEC policy to explode bombs only when winds were blowing away from Los Angeles and Las Vegas, towards the small towns of Nevada, Utah, and Arizona. A few hours after the test, AEC monitors found high levels of radioactivity in an area near the test site where, according to the Public Health Service, 1,400 people lived. Traffic was stopped on all major roads for two and a half hours. Government monitors checked 250 vehicles for radioactivity. Forty, including a Greyhound bus heading for Las Vegas with 30 passengers, had to be decontaminated. Though the fallout alert caused some unease among locals, most accepted the AEC's assurances that the levels of radioactivity, though higher than usual, had not been dangerous.

Gloria Gregerson was 13 years old during the Upshot-Knothole series. She lived in Bunkerville, Nevada, a small town about 100 miles downwind of the test site. She remembered being fascinated by the fallout. 'It doesn't snow where we live, and it was fun to pretend this was snow. We would shake the dust on to our heads and our bodies . . . Then I would go home and eat. If my mother caught me in time, I would wash my hands. I made most of the bread for our family from the time I was quite small.' When she was 17, Gregerson was diagnosed as having cancer of the ovaries. Later she developed cancer of the intestines and of the stomach, skin cancer, and leukaemia. After 13 major operations, she died in 1983, aged 42.

Many families continued their outings to watch the spectacular blasts. William Sleight of St George, Utah, a small town 135 miles due east of the test site, recorded one such excursion in his diary entry for 19 May 1953.

> Beautiful morning. We left St George at 4 a.m. for Las Vegas, Nevada. We were watching for the A-Bomb explosion on the

desert north of Las Vegas. At 5 a.m., just dawn, we saw the flash which lit up the skies, a beautiful red, visible for hundreds of miles away . . . After we came back on Highway 91, we were stopped and a young man examined our car with an instrument to see if we had picked up any radioactive dust while travelling on the Highway. Found none so we missed a free car wash . . . Returned to St George in a high wind which always seems to follow these explosions.

The shot that Sleight and his family witnessed was code-named Harry. It has since come to be known as 'Dirty Harry'. Soon after the blast, AEC technicians found that radioactivity readings in St George were very high – 6 rems per hour. At that rate, it would take only two and a half hours for residents to receive as much radiation as workers were then permitted to receive in a year. By the time Sleight got back to St George, local radio stations, on instructions from the AEC, were warning people to stay indoors, with their windows closed, until noon. Afterwards, an AEC spokesman said that radioactivity in St George was 'a little bit above normal, not in the range of being harmful'.

At Orderville, a little mining town 40 miles further east of the test site, five people allegedly developed symptoms of radiation sickness after Harry. The AEC sent a team of experts to Orderville to examine the men. The local newspaper reported that the commission emphasized 'that the mission was to reassure area residents that they could not possibly have been harmed by radioactive fallout'. Shortly after the explosion, 4,200 sheep, that had been grazing 50 miles north of the test site at the time of the blast, died from unknown causes. Some of the ranchers suspected fallout, but the AEC argued that fallout levels were too low to have killed the sheep. Malnutrition caused by an unusually hard winter was a more likely cause, the commission announced. As the *New York Times* commented in an editorial, this 'was not much consolation to the sheepmen, but by the same token implied reassurance to the population at large'.

The AEC was finding, however, that such reassurances were carrying less and less weight with downwinders. Recurring fallout alerts, reports of radiation injuries to people and animals, and fears about genetic damage were eroding their confidence in the commission. From late-1953 on, newspaper articles, especially in the *Deseret News*, the most respected and widely read of the local papers, grew increasingly concerned about fallout, and sceptical about the AEC's assurances that the tests were safe. In June 1953, Gordon Dunning, a top AEC staffer, told a commission meeting that 'the people in the vicinity of the Nevada Proving Ground no longer had faith in the AEC'. Test officials were frustrated by what they saw as a near-hysterical fear of fallout, which was being blamed for anything and everything that went wrong in the downwind areas. One disgusted AEC staffer told Daniel Lang that telephone callers asking about fallout levels often 'break down completely – crying and carrying on about what's going to become of the world'.

The AEC feared for the future of its domestic test programme. After the problems with Harry, commissioner Eugene Zuckert warned that 'In the present frame of mind of the public, it would take only a single illogical and unforeseeable incident to preclude holding any future tests in the United States.' The AEC took this possibility seriously. Testing stopped at Nevada for a year and a half while an AEC committee considered the effects of permanently closing down the domestic test programme. In 1954, after more than a year of study, the group concluded that testing in Nevada must continue and that, given a vigorous public relations campaign and some improvements in test procedures, it could continue.

But the worst was yet to come. For eight months after Upshot-Knothole ended, the US tested no atomic weapons. Then on 1 March 1954, on an islet near Bikini atoll, it exploded Bravo, the first full-scale hydrogen bomb. Hydrogen bombs, which use atomic bombs to trigger a nuclear-fusion reaction, are many times more powerful than ordinary atomic bombs. The first hydrogen bomb, an experimental device code-named Mike, was exploded in late 1952 on Enewetak, another Marshall Island atoll. Mike completely destroyed one island and left a one-mile gap in Enewetak's reef.

Despite last-minute weather readings indicating that 'winds at 20,000 feet were headed for Rongelap' – a small, but inhabited island about 90 miles to the east of Bikini – Bravo went ahead on schedule. The 86 Rongelap islanders were not evacuated or warned to take precautions, a fact which has led to charges that the authorities deliberately exposed them to fallout in order to obtain data on the health effects of fallout.

Bravo had the force of 15 million tons of TNT, one thousand times greater than the bomb that destroyed Hiroshima. It spewed radioactive debris many miles up into the stratosphere and thousands of square miles across the Pacific. Within four hours of the explosion, fallout from Bravo was settling on Rongelap. A fine white ash landed on the heads and bare arms of people standing in the open. It dissolved into water supplies and drifted into houses. The snow-like debris fell all day and into the evening, covering the ground up to an inch deep. On the day after the blast, three Americans wearing protective suits came to the island. They took readings from two wells with a Geiger counter and left after 20 minutes – without saying a word, according to the islanders.

Within 24 hours of the test, the people of Rongelap showed symptoms of acute radiation sickness, including nausea, vomiting, diarrhoea, and itching of the eyes. Almost all had burns on their skins. Many later lost all their hair. Two days after the test, the US evacuated the people of Rongelap to Kwajalein atoll, the command centre for the tests. The only public announcement of the emergency on Rongelap was a five-sentence press release from the AEC. Calling the explosion of the first deliverable hydrogen bomb 'a routine atomic test', the AEC stated that the evacuations were 'according to plans as a precautionary measure . . . There were no burns. All are reported well.'

Scientists at the Brookhaven National Laboratory calculate that each resident of Rongelap received from the blast a 'whole-body external radiation dose' of 175 rems, 25 times more radiation than members of the public are now allowed to receive in a lifetime. No one knows how much internal exposure the Rongelapese got from eating, breathing, and drinking fallout particles. Brookhaven National Laboratory still monitors the people of Rongelap. The 29 children who were under ten years old at the time of the blast grew more slowly than normal. The rate of thyroid cancer among the 10- to 18-year-old group was six times that in a similar group of children who were not exposed to fallout. Blood tests in 1964 revealed that some of the exposed population had broken chromosomes that could be attributed to radiation exposure. The local people say that a high proportion of pregnancies end in miscarriage and that many babies are born dead or so grotesquely malformed that they die within hours of birth. One mother described the child she bore after Bravo: 'It did not look human . . . it was like the innards of a beast.' The child was dead at birth. Two deaths on the tiny atoll – one from leukaemia, the other from stomach cancer – have been linked to fallout from Bravo. The mother of one of the dead boys later remarked to Australian filmmaker Dennis O'Rourke: 'Americans are educated people. Do they really think that one person's life is unimportant? They think they are smart . . . They are smart at doing stupid things.'

The US government returned the Rongelapese to their atoll in 1957, assuring them that radiation intensity had fallen to a safe level. Twenty-two years later, the US government announced that parts of Rongelap were still dangerously radioactive. The islanders asked to be moved to a new atoll, but the United States has refused, saying that there is no danger as long as people stay away from the northern islands and eat imported tinned food. Finally, in 1984 the islanders asked the environmental group, Greenpeace, for assistance. In 1985 Greenpeace helped the Rongelapese move to another atoll in the Marshall Island chain.

The outside world first learned of Bravo's disastrous effects two weeks after the blast when a Japanese tuna trawler, *Fukuryu Maru* (the *Lucky Dragon*), limped into its home port two weeks after the explosion. On the morning of Bravo, the *Lucky Dragon* had been at anchor 90 miles east of Bikini, 40 miles outside the official danger zone. Three hours after the blast, a fine white ash began falling on the boat and its crew of 23. The artificial snow storm continued for about five hours, laying down a mantle of ash so thick that the crew left footprints on the deck as they walked. Disturbed by this strange phenomenon, the fishermen decided to head for home. Before they left, the fishermen washed down their vessel, an act which probably saved many lives.

By the time the *Lucky Dragon* reached Japan, two weeks after the explosion, the entire crew was suffering from radiation sickness. Not knowing why they were sick, some of them simply went home. Others went directly to the hospital, where doctors quickly diagnosed their ailments and ordered that all 23 be brought in for observation.

News of the tragedy made headlines in Japan and touched off a tuna scare throughout the country. To reassure the public, the government was forced to check all fish catches for radiation. Inspectors destroyed one million pounds of contaminated fish. As the panic grew, so did anti-American feeling, expressed in letters to newspapers and violent street demonstrations. Seven months after the blast, the entire crew was still in hospital, receiving blood transfusions. At the end of September, a 39-year-old crew member died. The American ambassador to Japan sent the widow a cheque for $2,800 'as a token of the deep sympathy' of the American people. The US later paid $2 million compensation for the loss to the Japanese fishing industry.

Soon after the *Lucky Dragon* arrived in Japan, Lewis Strauss, a prominent Wall Street banker and the newly appointed chairman of the AEC, went to the Pacific on a 'fact-finding tour'. On his return to Washington at the end of March, Strauss was vigorous in defence of the test programme. He suggested that the *Lucky Dragon* was a 'Red spy outfit'. He denied that fallout from Bravo had contaminated fish catches. And he assured the American public that there would be no harmful fallout on the continental United States from tests far away in the Pacific. The AEC had already admitted that radiation levels in the United States had risen since Bravo, but Strauss said that they were 'far below levels which could be harmful in any way'.

Bravo, in Daniel Lang's phrase, 'was the shot that made the world fallout-conscious'. In November 1954 – eight months after Bravo – *Time* magazine reported that 'talk and worry over the H-bomb's radioactive "fallout" is spreading'. But in its report of the disastrous test, the AEC found one bright spot. Without the data from the contaminated islanders and fishermen, the report said, 'we would have been in ignorance of the effects of radioactive fallout and, therefore . . . much more vulnerable to the dangers from fallout in the event an enemy should resort to radiological warfare against us'.

In the United States, Bravo eventually faded from the headlines, but concern about fallout did not. In the late 1950s, downwinders began noticing what seemed to be unusually high rates of rare diseases, cancers and leukaemias. In 1980 St George resident Elmer Pickett looked back on those days for *Life* magazine: 'My father and I were both morticians, and when these cancer cases started coming in I had to go into my books to study how to do the embalming, cancers were so rare. In '56 and '57 all of a sudden they were coming in all the time. By 1960 it was a regular flood.'

Britain and the Soviet Union were also testing atomic bombs, but their tests were even more secret and free from public scrutiny than those of the Americans. Most of the almost 200 above-ground Soviet tests were conducted in Siberia, but even today little is known about the Russian test programme. Between 1952 and 1958, Britain conducted 21 major atomic tests and hundreds of smaller trials. Its testing ground was Australia and several South Pacific islands. The British government assured people that 'there will be no danger to the health of the people or animals in the Commonwealth', but many British and Australian soldiers were exposed to fallout, as were nomadic Aborigines who freely roamed the testing fields, unwarned and unprotected. Soldiers were threatened with execution if they said anything about the tests. An Australian government inquiry confirmed in 1985 that plutonium particles from the testing were, and still are, scattered over large areas of the Australian outback.

Whatever the effects of weapons testing on the Aborigines of Australia, South Pacific islanders, and Siberian peasant communities, it was the American testing that made the world think, and worry, about fallout. But there were many fallout incidents in the United States about which the public knew nothing. AEC officials knowingly released large amounts of plutonium into the air in various so-called 'safety tests' conducted at the Nevada Test Site from 1955 to 1958, and again in 1962. The purpose of the tests was to see what would happen if atomic bombs were attacked with conventional weapons. A 1974 AEC survey found high levels of plutonium in the soil over a large area of Utah and Nevada as a result of these experiments.

In the last two weeks of October 1958, the US exploded 20 atomic bombs in Nevada. Unusual weather meant that much of the fallout from these tests was dumped on Los Angeles, along with fallout from some earlier Soviet tests. For the week or so that the high levels persisted, the average person in Los Angeles received approximately the maximum annual permissible dose of radiation, but the Department of Health's advisory committee on radiation felt that the incident was more a public relations than a health problem. As Lauriston Taylor emphasized to the committee, the exposure standards were not precise. 'It is a tragedy that these things have gotten to having the effect of law,' he said. 'They carry the implication that we know what we are talking about when we set them. But, in actual fact, they really represent the best judgement we could exercise now, in the total absence of real knowledge as to whether they are correct or not.'

If, as Taylor believed, the standards are 'probably . . . way on the safe side of things', then the chief victim of the fallout situation was likely to be the future of the nuclear industry, not the health of the people of Los Angeles. The committee agreed that it was important not to alarm people unnecessarily. 'If you ever let these numbers get out to the public,' said Taylor, 'you've had it.' The geneticist Edward B. Lewis agreed, saying, 'This is certainly a confidential memo.' Lewis explained how he and his colleagues at the California Institute of Technology had outsmarted the journalists who had asked for information about the fallout incident. We 'held off long enough so they lost interest, because they had a three o'clock deadline'. The committee went on to discuss the importance of educating the public and the press about atomic radiation.

The AEC did everything in its power to preserve the embattled testing programme. It questioned the motives of those who sought to halt testing. It offered assurances about the quality of its safety programme. It said that fallout was not particularly hazardous, and that there was not much of it.

The psychology of the Cold War period in the United States, in which fear of a Communist take-over flourished and being 'soft' on defence was an indication of disloyalty, was an important weapon in support of atomic testing. In 1951 Senator Brien McMahon, a key figure in establishing the AEC and chairman of its congressional watchdog, the Joint Committee on Atomic Energy, stated that 'War with Russia is inevitable. We must blow them off the face of the earth quick, before they do the same to us – we haven't much time.' Anyone who questioned the safety of testing atomic weapons was opening himself or herself to accusations of disloyalty. As Karl Morgan, who was then director of Oak Ridge's Health-Physics Laboratory, remembered that period: 'it became unpatriotic and perhaps unscientific to suggest that atomic weapons testing might cause deaths throughout the world from fallout'.

In March 1955, David Lawrence, in his syndicated newspaper column, announced the existence of a Communist plot to undermine the US testing programme. 'Many persons are innocently being duped by it and some well-meaning scientists and other persons are playing the Communist game unwittingly by exaggerating the importance of radioactive substances known as "fallout".' Later that month Lawrence returned to the theme, 'The big bombs are not tested in this country, but in ocean areas far away from this continent. The Communist drive, however, is to stop all tests.' Also in March 1955, the headline of a column by Jack Lotto in the *Los Angeles Times* announced, 'On Your Guard: Reds Launch "Scare Drive" Against US Atomic Tests.'

At every opportunity, spokesmen for the AEC, and other official agencies, reiterated the message that the ill effects of fallout – if there were any – were greatly to be preferred to the prospect of a Communist take-over. When, in 1957, Albert Schweitzer, the doctor, missionary, and winner of the 1952 Nobel Peace Prize, called for an end to testing because of his concern about fallout, Willard Libby, an AEC commissioner and nuclear chemist, acknowledged that 'radioactivity may exceed safety limits in some places if peacetime tests are continued at the present rate'. But, Libby argued, this was a technical problem with technical solutions, such as replacing contaminated milk with imported supplies, whereas inadequate national defence could be fatal. 'I ask you,' Libby said in his public response to Schweitzer, 'to weigh this risk against what I believe to be the far greater risk – to freedom-loving people everywhere in the world – of not maintaining our defenses against the totalitarian forces at large in the world.'

To the downwinders, the message was straightforward. A widely distributed AEC pamphlet dated March 1957 advised 'you people who live near the Nevada Test Site' that 'your best action is not to be worried about fallout'. During the Plumbob series, in the summer of 1957, the commission announced, 'we can expect many reports that "geiger counters were going crazy here today". Reports like this may worry people unnecessarily. Don't let them bother you.'

The AEC was a notoriously secretive organization. Its monopoly of information about testing made equal intercourse with outsiders impossible. Even Senator Brien McMahon, who was an enthusiastic proponent of the test programme, criticized the commission's suppression of information. In 1949, he complained that although Congress had appropriated $3 billion for atomic weapons, not even the Joint Committee on Atomic Energy knew how many bombs the United States had. JCAE chairman Representative Chet Holifield, a strong supporter of atomic weapons and atomic energy, complained that 'we [have to] literally squeeze information out of the agency', and charged the AEC with 'selective use and release of information to favor a political position'. President Eisenhower himself encouraged the agency to exploit the public's ignorance on atomic matters. After meeting with Eisenhower on 27 May 1953, to discuss the fallout problem, AEC Gordon Dean chairman, wrote in his diary, 'the President says, "Keep them confused as to fission and fusion."'

After Bravo, the AEC began releasing some of its fallout data. Realizing that as far as health matters were concerned, the public had more confidence in the Public Health Service than the AEC, the commission in 1954 assigned the Health Service to monitor fallout. Twenty-five years later, a special presidential investigator found that the AEC had imposed strict and secret conditions on the Health Service's work. Chief among them was that 'no public statement could be made by the PHS with respect to radiation issues and that all press releases had to be cleared with the AEC and the White House'.

Some insiders, particularly military men, believed that the AEC's troubles sprang from the fact that it was not secretive enough. The commission, they said, was scaring people by dwelling on the dangers of testing. At a meeting with the commission in mid-1953, the chairman of the Pentagon's Military Liaison Committee warned that 'arousing public fears' would lead to 'large-scale public opposition to the continental tests'. Other participants agreed, saying that 'precautions taken by the AEC ... caused undue public concern', and accusing the AEC of 'over-emphasizing the effects of fallout'. Commissioner Henry Smythe disagreed. Reaffirming 'the extreme care taken by the AEC in avoiding hazard to the public is believed to be the best way to create confidence and allay public fear,' he said. But, according to Robert Sherwood, the Pulitzer Prize-winning playwright and historian, and wartime speechwriter for Franklin Roosevelt, the AEC operated under the theory 'that the way to keep people fearless is to keep them ignorant'.

The deep secrecy surrounding the tests extended even to the radiation exposure limits adopted by the AEC. The commission was trying to build public confidence in a set of safety standards it would not disclose. Even radiation specialists capable of discussing the technical issues involved were excluded from the debate. Public confidence was thus to be a matter of faith, rather than reason. In a 1953 editorial the *Deseret News* remarked helplessly that 'the AEC assures us that they [levels of radiation] have been "well within the limits of safety" – whatever those are'.

In exposing workers to radiation, the AEC generally followed the recommendations of the National Council for Radiation Protection. But there were some important discrepancies between the policies and practices of the AEC and those recommended by the NCRP. In 1948, the NCRP had given notice that it planned to cut the permissible external exposure limit for workers from 0.1 rem per day to 0.3 rem per week. In 1950, the AEC adopted the lower figure for workers in its weapons plants, but the standard was not lowered for test workers until after the first Nevada Test Series, when it became clear that such a move would not hamper the test programme. The AEC also had a secret provision allowing higher exposures to test workers in emergencies. As a classified report put it, 'The [AEC's] Division of Biology and Medicine agreed to an unpublicized exposure of 3.9 r ... where absolutely necessary.'

Perhaps the most important difference between the AEC and the NCRP concerned the question of whether there is a safe level of radiation – a threshold below which radiation causes no damage. In 1948, the NCRP had adopted the 'no-threshold' concept, the idea that there is no safe level of radiation. The NCRP's guiding principle of radiation protection then, as now, was explained in 1956 by Lauriston Taylor, the council's longest-serving member: 'Any radiation exposure received by man must be accepted as harmful. Therefore, the objective should be to keep man's exposure as low as possible and yet, at the same time, not discontinue the use of radiation altogether.' Thus the NCRP's exposure limits could not and did not promise absolute safety. AEC officials were aware of this, as a secret memo written in 1953 by John Bugher, Shields Warren's successor as director of the commission's Division of Biology and Medicine shows. In the memo Bugher said that he knew of 'no threshold to significant injury in this field'. In the same year, however, the AEC published a pamphlet saying that 'low-level exposure can be continued indefinitely without any detectable bodily change'. For several decades, AEC officials continued to publicly assert that there was a threshold of safety and that its exposure limits were below that threshold.

Protecting the public from fallout presented many difficult practical and theoretical problems. Detecting and measuring radiation over a large area was expensive, time-consuming and labour-intensive. If levels did get too high in some areas, it was not possible, as it would be in the workplace, to send people home or instruct them to wear protective clothing. More drastic measures, such as ordering a mass evacuation, might lead to panic – and undermine confidence in the test programme.

Officials were thus wary of adopting a hard-and-fast standard for the public, one that if exceeded would trigger stringent protective measures or evacuation. Acknowledging 'the somewhat delicate public-relations aspects of the affair', Thomas Shipman, the physician in charge of RadSafe for the first Nevada tests, decided not to set a standard for public exposure for the first series. He was against evacuation unless exposures were likely to exceed 50 rem (500 times the amount now allowed to the public annually) over a four-week period, though even then 'such a step should not be taken unless it is firmly regarded as essential'.

Pressure from an *ad hoc* group of government radiation experts called the Feasibility Committee forced Shipman to reduce exposure limits for the public during Buster-Jangle, the second test series. The experts, led by Shields Warren, insisted that the public should not be allowed to receive more radiation than radiation workers. Specifically, the committee wanted to extend to the public the workers' limit of 0.3 rem per week (15 rem per year), which the committee called 'the only generally recognized safe maximum permissible dose'. The committee's advice clashed with that of an AEC panel, which suggested that, because the public was not subject to repeated regular exposure, 'the allowable dose can, from the safety point of view, very well be raised to five or ten roentgens (publicity considerations disregarded)'. Finally, however, the test authorities agreed to extend the occupational standard to the public.

Two years later, in 1953, some AEC staffers argued that the public standards should be lowered even further. The commission, they said, 'has no right to subject the public to the radiation permitted for its own employees'. Their reasoning was that employees freely choose to accept hazardous conditions, are paid for their work, and are closely monitored for radiation exposure – none of which conditions apply to members of the public who happen to live near the test site or in the path of fallout. In 1955 John Bugher reduced the public limit from 15 to 3.9 rem per year, on 'the principle that an exposure level which may be acceptable occupationally should be reduced by an appreciable factor where large populations are concerned'. If any amount of exposure carries a risk in proportion to its size, then many small exposures could produce even more damage than a few large exposures. Bugher was particularly concerned about genetic damage, which if widespread enough could swamp future generations. However, the AEC's new limit was still significantly higher than the limit of 1.5 rem per year that the ICRP had adopted in 1954 for members of the public.

The most efficient way of controlling radiation exposure to the public was to reduce it at its source – that is to produce less fallout. This was technically possible, AEC staffer Gordon Dunning told a congressional committee in 1957. However, he cautioned, 'If we continue to reduce the fraction we are willing to release, we eventually reach a cost of control which makes the operation prohibitive. The dilemma is that we must weigh the degree of undesirability of radioactive fallout against the advantages which may be anticipated from activities which are inevitably accomplished by fallout.' 'Our great hazard is that this great benefit to mankind will be killed aborning by unnecessary regulation,' Willard Libby warned his fellow commissioners. A law passed in 1946 enabled the government to ignore one potentially ruinous cost, damage claims filed by people alleging radiation injuries from the tests. The courts have ruled by a 1946 law prohibiting citizens from suing the US government for injuries arising from the execution of its policies, even if negligence was involved.

How much to pay for safety was a perennially thorny issue for the AEC, which was always under pressure to keep safety costs down. The commission's advisory Committee on Biology and Medicine concluded at a secret meeting in October 1953, that the civilian contractors who ran the AEC facilities were spending too much money on protection – indulging in superfluous safety measures. The problem, committee-member Gioacchino Failla explained, was that 'on general principles, the contractor wants to be on the safe side, so the expense to the AEC is quite appreciable'. Meanwhile the Bureau of the Budget was asking 'what was being done by the AEC to reduce the cost of health protection'. The answer, the committee felt, was to adopt a new policy that would require contractors to cut back on health and safety expenditures. Failla supported the policy, but feared that 'if this information or policy should be released to the public, it would create a rather bad impression'.

CHAPTER TWELVE

Fallout

In the early years of the test programme, fallout was largely a mystery. In some respects, it remained mysterious to the AEC longer than it did to others. The AEC's efforts to improve its scientific understanding of fallout were hampered by its military-political obligations. The commission's obsession with secrecy stifled a free exchange of information and ideas. Its commitment to testing discouraged the pursuit of information that might indicate that testing was hazardous. Scientists outside the AEC were free from these pressures and, despite their lack of access to important data, made many important discoveries about fallout. The geneticist Edward B. Lewis was the first to suggest that iodine–131 was a hazard to the thyroid glands of children. The Nobel Prize-winning chemist Linus Pauling was the first to show that carbon–14 is hazardous to humans. Harvard biochemist Herman Kalckar was the first to suggest a way to measure levels of strontium–90 in children. Scientists from Montana and Utah were the first to show that there were high concentrations of fallout downwind of the Nevada Test Site.

What was known about fallout in the 1950s? The very first atomic explosion, the Trinity test, had confirmed suspicions that atomic explosions produce radioactive debris that later falls back to earth. The concept of fallout remained obscure, however. The word itself did not come into common use until around 1952. Until then writers used cumbersome phrases, such as Bradley's 'dissemination of radioactive materials by air and water' or, more emotively, 'the evil cloud'.

The discovery that Trinity had contaminated wheat in Indiana showed also that fallout could travel thousands of miles before returning to earth. *Time* magazine asked, 'How far do atomic bombs spread their radioactive end-products? This question has never been answered conclusively . . . Some scientists believe that active particles from a single bomb are carried all over the world.' In fact, even before the Bravo explosion raised radiation levels around the world, the AEC had assigned US government employees stationed overseas to collect rain and dust for radiation analysis. The commission also imported foreign cadavers for radioanalysis in an attempt to keep tabs on worldwide radiation levels.

Until at least the early 1960s, the AEC operated on the assumption that fallout would be distributed uniformly around the world. But from the very beginning of the test programme, it was clear that there were 'hot spots' – areas of concentrated radioactivity. Wind patterns kept most of the fallout in the Northern Temperate Zone, and rain and snow storms created high concentrations of fallout in places many thousands of miles away from the test site.

Two days after the first atomic explosion in Nevada, in January 1951, a heavy snow storm hit Rochester, New York. At the Eastman Kodak plant in the city, Geiger counters began to go wild. Readings were 25 times higher than normal. Kodak executives realized at once what had happened: an atomic test had made the snow radioactive. And, knowing the damage the Trinity fallout had done to their film, they were alarmed. On the company's behalf, the National Association of Photographic Manufacturers sent the AEC an urgent cable: 'Situation serious. What are you doing?' The AEC could not assure the film manufacturers that there would be no more fallout incidents, but to forestall possible lawsuits, officials agreed to provide Kodak and other photographic companies with advance information about all atomic tests, including maps indicating the likely fallout path. Several businessmen were given security clearances so that they could receive this classified information, which was denied to the public.

After the Kodak incident, the AEC began to monitor the path of fallout clouds as they crossed the country until they dispersed or reached the Canadian border. Publicly, the commission continued to downplay the chances that testing would contaminate any area except the test site. But, on 27 April 1953, something happened at the Rensselaer Polytechnic Institute in Troy, New York, to change that position. As Professor Herbert Clark's radiochemistry students gathered for their Monday morning lab session, Clark noticed that all the Geiger counters were registering high levels of radiation. Some of the students took the counters outside, and obtained readings as much as one thousand times higher than normal. Clark immediately guessed that the high activity was due to fallout washed down in the previous night's rains. He telephoned John Harley, a friend who worked for the AEC's Health and Safety Laboratory in New York, and told him about the readings his students were finding. Thinking that Clark's story was a practical joke, Harley hung up. He soon called back, however, and informed Clark that the high readings were probably due to fallout from a recent test in the troublesome Upshot-Knothole series. Test Simon was detonated on 25 April. Within a few hours, it had dropped high levels of fallout on communities near the test site in Nevada. Simon's fallout cloud then passed over Utah, Colorado, Kansas, Missouri, Illinois, Indiana, Ohio, and Pennsylvania, dropping radioactive debris as it went. Finally, over upstate New York and nearby parts of Vermont and Massachusetts, a massive thunderstorm washed the cloud out of the sky.

Clark's students fanned out to neighbouring towns and tested sidewalks, streets, roofs, plants, and puddles for radioactivity. Readings were generally 20 to 100 times normal, though some were even higher. 'The activity of drinking water was . . . about 100 to 1,000 times greater than the natural radioactivity,' Clark reported in *Science* magazine. Many porous or rough-textured items had collected so much radiation that they could produce X-rays of themselves, called 'auto-radiographs'. Radiation levels declined fairly rapidly, however, and the AEC decided that there was no need to filter tap water or try to decontaminate affected surfaces, such as pavements and roofs. This was just as well, since several attempts to clean contaminated objects by determined scrubbing and the use of concentrated hydrochloric acid failed. Clark called the fallout levels 'exceptionally high, though not hazardous'.

Not until after this incident did the commission acknowledge the need to measure fallout on the ground rather than just in the air. But there still was a grave omission in the commission's fallout monitoring programme. As far as the AEC was concerned only the external radiation given off by fallout was important. AEC officials ignored the hazards of internal exposure to fallout, even though internal exposure to some fission products is very hazardous. The terrible effects of inhaling or swallowing radioisotopes had been well known since the 1920s from studies of the New Jersey dial painters and other radium victims. Nonetheless, for many years, the commission monitored only gamma radiation, not even attempting to regulate or measure how much fallout people were breathing or taking in through the food chain.

The AEC's own studies had highlighted the existence in fallout of substances that would be hazardous internally. Project Gabriel, a secret investigation of the health effects of fallout begun shortly after the war, found in 1949 that 'there is little question but what there is real danger to inhaling particulate radioactive substances in such finely divided particles as to be retained within the lung'. The commission's official line was that though internal exposure to fallout was theoretically possible, it was unlikely actually to occur.

The AEC knew, for example, that strontium–90 – a radioisotope that is only produced in atomic explosions – concentrates in bones and can cause bone and blood cancers, but it did not try to measure the strontium–90 intake of people exposed to fallout, because, it said, strontium was not likely to get into the body. Some outside scientists said that people could ingest strontium–90 by drinking the milk of cows that have eaten contaminated plants, but the commission pooh-poohed this notion. In its 1953 report to Congress, the AEC stated that, 'the only potential hazard to human beings would be the ingestion of bone splinters . . . As to the taking in of radioactivity by animals eating plants grown in soil affected by the fallout from the tests, experiments have indicated that there is no hazard to human health from this source.' In private, however, the AEC had already identified strontium as the most worrisome element in fallout. In 1953 it commissioned the Rand Corporation to conduct a secret survey of strontium contamination throughout the world. The study, called Project Sunshine, measured strontium contamination in 'sunshine units'.

Strontium–90 was not the only constituent of fallout of which the AEC was, or appeared to be, ignorant. For some years, the AEC had ignored radioactive iodine–131, a component of fallout that concentrates in the thyroid and can cause thyroid cancer. By 1954 the commission was aware that fallout contains radioactive iodine and that it could contaminate milk. But it did not begin to systematically monitor fallout in the food chain until 1958, and then, as Charles Dunham, director of the Division of Biology and Medicine, later said, 'we were too busy chasing strontium–90', to pay attention to iodine–131.

Its own ignorance about fallout – and that of the public – enabled the commission to make misleading statements about the safety of testing. AEC officials frequently compared exposure to fallout to having a chest X-ray or wearing a watch with a radium dial, which it said exposed the wearer to at least eight times more radiation than did testing. Such comparisons are inappropriate because they take account only of the external radiation exposure, and ignore the more hazardous internal exposure which is a risk of fallout, but not of chest X-rays or wearing a watch.

In an article in the 25 March 1955 issue of *US News and World Report*, Libby said that the AEC had evidence that fallout was 'not likely [to] be at all dangerous'. A number of prominent scientists, among them Linus Pauling and geneticists Hermann Muller and Curt Stern, wrote to the commission, protesting that Libby's statement would 'increase the existing distrust by many thoughtful people and particularly by scientists of the intrinsic reliability of public information coming from the AEC'.

Declassified minutes of an AEC meeting in February 1955 reveal more concern about fallout than was evident in the commission's public statements. At the meeting, held the day after a test in the troubled Teapot series, Libby took his usual robust view of the matter, declaring that 'People have got to learn to live with the facts of life, and part of the facts of life are fallout.' To this, Chairman Lewis Strauss replied, 'It is certainly all right they say if you don't live next door to it.' 'Or live under it,' commented another commissioner. Later in the same meeting, there was a discussion about the least injurious path for the fallout clouds to travel. 'Isn't it east?' asked a commissioner. 'No,' said Strauss. 'East, they go ... over St George, which they apparently always plaster.'

From the mid-1950s on, the scientific debate in the United States about fallout and the effects of low-level radiation became ever more polarized. Scientists began to be identified as either for or against the AEC. Political and moral issues were intertwined with the technical ones, and many if not all the scientists involved in the debate felt as strongly about the non-scientific questions as they did about the scientific ones. At the heart of the debate were two very different hazards to society – the hazard of disease and genetic damage and the hazard of totalitarianism. In all this, the general public was at a loss, ill equipped to participate in or even to referee a debate in which technical and political issues were so inextricably linked. Senator Anderson summed up the layman's frustration: 'You get one group of scientists together, and they say one thing, and you get another group together, and they say another thing. What does a man who is not a scientist have that he can tie to?'

In 1955 an embarrassing storm blew up over the AEC's suppression of a paper by the Nobel Prize-winning geneticist Hermann Muller. In May of that year the commission accepted an abstract of Muller's paper for a United Nations-sponsored conference on the 'peaceful uses of the atom' to be held in Geneva later that year. The paper was to assess worldwide fallout exposure and the consequent genetic damage. In July, when he tried to submit his full paper, however, Muller was told that the UN had dropped him as a speaker because the programme was too crowded. Muller was also informed that he could not take part in the debate at the conference, because there was no space for him in the US delegation. After the *Washington Post* reported that UN officials denied having blocked Muller's paper, the AEC admitted that it, not the UN, had been responsible. The reason, officials said, was that Muller's paper mentioned Hiroshima, which was unsuitable at a conference on the peaceful use of atoms. After the conference, the respected journal *Science* ran an editorial called 'Liquidating Unpopular Opinion', accusing Strauss of politically motivated censorship. 'Chairman Strauss has consistently maintained that fallout from tests of nuclear weapons have been so low that they could not bring harm to human beings. Muller has repeatedly presented reasons for believing such complacency to be unjustified ... could it be that Muller's persistence in disagreeing with the chairman of the commission was a factor in barring his report?'

During the spring and summer of 1957, the US Congress's Joint Committee on Atomic Energy held hearings on the nature and effects of fallout. The hearings focused more on the AEC's handling of the fallout problem than on the fallout itself. A number of witnesses criticized the commission for consistently underplaying the risks of radiation exposure. Hermann Muller told the committee that, as the truth became known, the AEC's tendency to deny the risks of radiation exposure was weakening public confidence in the AEC. 'The only defensible or effective course for our democratic society is to recognize the truth, to admit the damage, and to base our case for continuance of the tests on a weighing of the alternative consequences,' he told the committee. The AEC still denied that there was a health problem. Charles Dunham, the commission's new Director of Biology and Medicine, told the committee that 'there is no anticipated danger of fallout from the [testing] programme'.

Other witnesses challenged the correctness of comparing the risks of fallout to those of other activities, like driving an automobile or climbing a ladder. Risks that are voluntarily chosen, they said, cannot be directly compared to those over which people have no control, such as exposure to worldwide fallout. AEC's dual role as weapons tester and assessor of fallout hazards was said to be in conflict, and to undermine confidence in the commission's scientific findings. After the hearings ended, committee chairman Chet Holifield wrote an article in the *Saturday Review* in which he was scathing about the commission: 'I believe from our hearings that the Atomic Energy Commission approach to the hazards from bomb test fallout seems to add up to a party line – "play it down".' The hearings did not settle any scientific questions, but they did undermine the authority of the AEC, and deepen the divisions that were appearing in the scientific community.

In 1958 the United States, the Soviet Union, and Great Britain tested almost one hundred nuclear weapons – more than twice as many as were ever exploded before in a single year. Two-thirds of the tests were conducted by the Americans, one-quarter by the USSR, and five by Great Britain. In November of that year, Russia and the United States agreed to a provisional test-ban. AEC official Charles Dunham, who in 1957 had told the JCAE that the test programme did not pose a fallout problem, later told President Kennedy that, in his opinion, if weapons testing had continued at the 1958 rate, 'civilized man would have been in trouble'.

For almost three years, the superpowers exploded no atomic bombs, but the earlier tests had left their legacy. In 1959, with no bombs being detonated, fallout levels reached a new peak. Record levels of strontium–90 were found in milk, wheat, and soil in parts of the United States. St Louis, Atlanta, and Mandan, North Dakota had high readings of strontium–90 in milk. There was seven times more strontium–90 in New York City's soil than there had been in 1954. Strontium–90 concentrations in Pittsburgh and Seattle had risen five times in one year. Samples of wheat in Minnesota and the Dakotas had fifty per cent more strontium–90 than the AEC had said was safe for human consumption. Willard Libby announced that the AEC was 'very concerned' about strontium contamination of wheat, which he blamed on tests that the Soviet Union had conducted the previous fall.

In March 1959, a new furore erupted when Senator Clinton Anderson released a classified letter in which the Department of Defense informed Libby that stratospheric fallout returns to earth in two years, not, as the AEC had maintained, seven. The less time fallout remains aloft, the more radioactive it is when it returns to earth. Libby said that he had kept silent about the new finding because it was classified, because its implications were not particularly significant, and because he disagreed with it. Nonetheless, the disclosure embarrassed the AEC. 'Those who have been worrying about the dangers of radioactive fallout, and charging the administration with suppressing and distorting the unpleasant facts seem to have been more right than their critics,' said the editors of *Commonweal.* Even the staid *Saturday Evening Post* ran a two-part series titled, 'Fallout: The Silent Killer', charging among other things that the AEC had under-estimated the danger from fallout.

As the public and politicians lost confidence in the AEC, there were suggestions that radiation protection should be shifted from the commission to the Public Health Service. Budget officials opposed such a shift, because 'the whole future of the use of radiation would depend on the decisions of officials whose major mission and experience is public health'. Congressmen and administration officials were also worried about the fact that US government agencies were being guided on such sensitive matters by the NCRP and ICRP, private – and in the case of the ICRP, largely foreign – groups.

Perhaps because some of its members came from countries with no nuclear testing programme, the ICRP seemed more inclined that the NCRP to maintain strict exposure limits. As strontium levels in milk were getting uncomfortably high, the NCRP began asking itself whether its existing limits might not be too stringent. 'It is the responsibility of the NCRP to make certain that the levels recommended for strontium–90 are realistic,' read the minutes of a 1958 meeting. The following year, the NCRP doubled the amount of strontium–90 workers could receive, and raised the amount allowed in milk by 25 per cent. One member said that 'the nation's security may demand the exposure of people to higher levels of radiation than those just established by the International Commission'.

In August 1959, Eisenhower announced the formation of the Federal Radiation Council (FRC) 'to advise the President with respect to radiation matters', and to reassure the public that radiation protection standards would be established objectively. Among the council's six members were the heads of the two agencies most affected by radiation regulations – the AEC and the Department of Defense.

The FRC issued guidelines on external radiation exposure in 1960, and on internal emitters the following year. The council's recommendations had no legal force: other government agencies were free to follow or ignore the council's advice, as they pleased. In general the council followed the NCRP's lead, but in 1961 it recommended that the NCRP limits on strontium–90 and radium–222 – already several times more lax than those of the ICRP – be raised by a factor of three. The FRC's move alarmed some scientists, but with testing temporarily halted and levels of strontium–90 declining, the issue of fallout was receding from the headlines.

Then, in September 1961, at the height of the Berlin crisis, the Soviet Union resumed testing. The United States and Great Britain followed shortly, and France joined in by exploding its first atomic bomb. More and bigger bombs were exploded in 1961 and 1962 than ever before, producing more fallout than ever before. Soon after the USSR resumed testing, the US Public Health Service found dramatic increases in iodine–131 concentrations in milk. In Des Moines, Minneapolis, Detroit, Kansas City, and Palmer, Alaska, levels of iodine–131 in milk approached the FRC limits. Levels in several other cities exceeded the FRC limits.

Invisible contamination of milk was understandably alarming to the public and to the dairy industry. In an effort to create confidence in milk supplies, President Kennedy announced at a 1962 conference on milk and nutrition that milk was served at every White House meal. Dairy farmers in Utah agreed to turn the contaminated milk into cheese and butter, thus allowing the radioactive iodine (which has an eight-day half-life) time to decay. When Soviet tests rained iodine–131 on Minnesota, the state offered its milk producers incentives to remove cattle from grazing and feed them stored forage.

The FRC objected to these counter-measures, calling them 'premature', even though its own guidelines for radioactivity in milk had been exceeded. The FRC's chairman announced that the council's guidelines did not apply to fallout, only to radioactivity caused by 'normal peacetime operations'. The council also stated that 'iodine–131 doses from weapons testing conducted through 1962 have not caused an undue risk to health'. The FRC was severely criticized for not responding to concern over fallout, which is what had led to its creation in the first place. In an attempt to silence its critics, the council considered saying that decontamination measures would be appropriate when there was ten times more iodine–131 in milk than its own guidelines allowed, but when the AEC said that even this was too restrictive the council backed down.

The AEC's fierce opposition to tightening the guidelines on iodine concealed dissent within the commission staff. In 1962, an AEC scientist named Harold Knapp studied the exposure of young children to iodine in milk, and concluded that existing standards were at least ten times too lax. His recommendation that the standards be tightened evoked this response from the director of the commission's Division of Operational Safety:

> The present guides have, in general, been adequate to permit the continuance of nuclear weapons testing, and at the same time have been accepted by the public, principally because of an extensive public information programme. To change the guides would . . . raise questions in the public mind as to the validity of the past guides.

In mid-1963, the United States, the Soviet Union, and Great Britain agreed in principle on an end to atmospheric testing. Since the first bomb was detonated in the New Mexican desert, almost 400 atomic bombs had been exploded above ground, releasing more than 200 million tons of radioactive fission products to the environment. People are still being exposed to the longer-lived components of the fallout from those explosions, including caesium–137, strontium–90, carbon–14, and various isotopes of plutonium – in rain, in food, in air and water. Much of the fallout was distributed globally, so no part of the world is free from it, though the northern hemisphere is more contaminated than the southern hemisphere. The body of every man, woman, and child on earth now contains some strontium–90, a substance that does not exist in nature. Each person is estimated to get a radiation dose of 4 to 5 millirems per year from this fallout.

In urging the Senate to ratify the test-ban treaty, Kennedy spoke of the risks of fallout: 'The number of children and grandchildren with cancer in their bones, with leukemia in their blood, or with poison in their lungs might seem statistically small to some, in comparison with natural hazards, but this is not a natural health hazard – and it is not a statistical issue. The loss of even one human life, or malformation of one baby – who may be born long after we are gone – should be of concern to us all. Our children and grandchildren are not merely statistics towards which we can be indifferent.' On 10 October 1963, the Limited Nuclear Test Ban took effect. With most testing gone underground (China and France continued to explode atomic bombs above ground), the fallout panic subsided, leaving in its wake a confused and cynical public.

PART THREE

CHAPTER THIRTEEN

1956 Standards

In 1955, hoping to demonstrate that the weight of scientific opinion was with the AEC, the commission's chairman, Lewis Strauss, asked the National Academy of Sciences to study the effects of radioactive fallout. The Rockefeller Foundation agreed to fund the project, and the academy established the Biological Effects of Atomic Radiations Committee (the BEAR committee) to prepare a report.

The BEAR committee's report was published in June 1956 simultaneously with a similar study by Britain's Medical Research Council. The two reports contained good and bad news for the AEC – and the public. Fallout, they concluded, was not currently at a dangerous level. Averaged over the entire population, medical X-rays were a greater source of exposure than fallout – and so might atomic power plants and radioactive wastes one day be. Nonetheless, fallout should not be allowed to rise to 'more serious levels', said the reports, because the genetic consequences would be tragic. People exposed to low-level radiation could unknowingly suffer genetic damage that would be passed on to their descendents. Because fallout travels around the globe, the whole world population was at risk of exposure. Though the number of injuries might be small in percentage terms, the absolute number could be very large.

Both reports recommended that all workers should be limited to an accumulated exposure of 50 rems up to the age of 30, and another 50 rems between ages 30 and 40. In practical terms, assuming a working life that begins at age 20, this meant that workers should be limited to 5 rems per year.

The BEAR committee believed, in the words of Warren Weaver of the Rockefeller Foundation, that 'any radiation in addition to the inescapable natural background is unfortunate and harmful from a genetic point of view', but it did not want to ban all extra exposure, since that would mean an end to medical and military use of radiation. The committee had considered recommending that annual exposures for workers be cut to somewhere between 3 rems and three-quarters of a rem but a survey of AEC plant managers convinced the committee that anything under 5 rems would be impracticable.

The world of radiation protection was a small one, and there was much overlap between the BEAR committee, the Medical Research Council, the NCRP and the ICRP. Both the ICRP and the NCRP decided to go along with the BEAR committee recommendations, for the sake of uniformity. ICRP, in fact, decided to recommend a 5 rems per year limit even before the BEAR and MRC reports were published. The NCRP followed suit in early 1957. Both the ICRP and the NCRP added numerous supplemental guidelines designed to accommodate occasional overruns of the new, stricter standards. Workers could receive up to 12 rems per year, as long as they averaged no more than 5 rems per year over their working lives.

The NCRP emphasized that the lowering of its exposure limits did not mean that the higher ones had resulted in injury. The change, said the committee, was simply 'based on the desire to bring the [limits] into accord with the trends of scientific opinion'. These new standards, like all the earlier ones, were not the result of precise calculations. They were compromises, estimates, and evolutions from earlier figures.

Before the war, it had been scarcely conceivable that the general public would ever need protection from radiation – which was then a highly specialized tool confined to the hospital and the laboratory. Afterwards, however, it was evident to many that, as one ICRP member said, 'ionizing radiations will be so common that there will arise protection problems on a scale never before conceivable and of importance not only for the individual but also for future generations'. In fact, in the absence of any limits on how much radiation could be dumped on the public, that sort of large-scale contamination had already begun.

For 12 years after its inauguration in 1944, the government's plutonium-production facility at Hanford in Washington State released radioactive wastes to the environment 'on a scale that today would be considered a major nuclear accident', as the *New York Times* put it more than 40 years later when the news became public. In 1945 alone, Hanford officials secretly released 340,000 curies of radioactive iodine to the surrounding countryside. By comparison, official estimates give the total release from the Three Mile Island accident as 15 curies. To a certain extent, the releases were the result of an inadequate filtration system, but some were 'planned experiments'. In one such experiment, carried out in December 1949, 5,500 curies of radioactive iodine was released over eastern Washington and Oregon. Researchers took radiation measurements, but made no attempt to warn civilians or to evaluate the health effects of the releases.

When the idea of setting formal limits on public exposure was first broached, there was opposition from several quarters. One of those quarters was the US Atomic Energy Commission. Commissioner Robert Bacher told Lauriston Taylor that such a limit would be 'psychologically dangerous', because 'the populace would think that any exposure would be harmful'. What Bacher wanted was a pro-gramme to convince the public that a certain amount of radiation was 'tolerable'. Taylor agreed that the public needed to be educated, but wondered how to achieve a balance between fear and indifference in the public's attitude toward radiation.

The first move to limit the public's exposure to radiation came at the 1949 Tripartite Conference, a high-level meeting of radiation specialists from the United States, Britain, and Canada. Experts there were worried that increasing the radiation exposure of large numbers of people, even by relatively small amounts, might result in genetic damage on a scale that could threaten the survival of the species. The Conference recommended that the radiation dose to the public should not be more than one per cent of the occupational dose.

The US delegation to the Tripartite Conference was not happy about the one per cent limit. Gioacchino Failla, a participant in the Tripartite Conference and a member of both the ICRP and its US counterpart, the NCRP, felt that the one per cent limit was unneces-sarily restrictive. In 1953, Failla and others succeeded in getting the conference to raise its recommended limit for public exposure from one per cent to ten per cent of the occupational dose. The following year, the ICRP also recommended that doses to members of the public be kept to ten per cent of the occupational dose.

But the Americans were still dissatisfied. They argued that creating a double standard would initiate a never-ending see-saw process in which workers would seek to eliminate the differential between them-selves and the rest of the population, and members of the public would fight to have it reinstated – in the end reducing permitted exposures to impractically low levels. Some members of the NCRP were so reluc-tant to introduce double standards that the council toyed with the idea of lowering the occupational dose limit by a factor of ten, so as 'to make this uniform for all categories rather than have the two standards'. The NCRP also argued that not enough was known about the genetic effects of radiation for a sensible limit for the public to be set. It did adopt the ten per cent rule, but hedged the issue for a few years, saying that it applied only to children. Finally, in 1956, the NCRP formally followed the ICRP and extended the ten per cent limit to individual members of the public. The AEC adopted the public limit in 1957, and Hanford's massive radiation releases came to an end.

In 1959, several years after having established different standards for workers and non-workers, the NCRP set up a special committee to review the justification for having done so. Fair or not, double standards were regarded as essential because exposing the whole population to ionizing radiation at the levels necessary to carry out radiation work might jeopardize society's ability to reproduce and survive. Fortunately, the committee identified a number of good reasons for allowing workers more radiation than the population as a whole, the chief ones being that employment in the nuclear industry is a matter of choice, that workers are informed of the risks of their exposures and paid for taking them, that the workforce is screened to eliminate those who may be unusually vulnerable to radiation, and that workers are closely supervised and monitored for radiation exposure.

Having decided that it was fair to set a stricter dose limit for the general poulation than for workers, the next question was, what should the limit be? In the absence of detailed scientific data for the new limit, both NCRP and ICRP liked the idea of allowing the population to get the same dose from man-made radiation as it was already getting from natural background radiation, on the grounds that 'this was a level to which the human population has been exposed for many generations while continuing to survive and flourish'. There were, however, some doubts about this approach: an NCRP report noted that 'there is hesitation in suggesting that an additional hazard be accepted on the basis that other hazards already exist'. Nevertheless, both groups ultimately decided that natural radiation was the best yardstick to use in limiting radiation from human activities.

The ICRP felt that the average dose of radiation from all sources except background should be limited to 'an amount of the order of natural background' – about 100 millirems per year. The commission also wanted the population dose limit to apply to all sources of non-background radiation, including medical radiation. The NCRP, however, wanted to allow people to receive *twice* as much radiation from human activities as they did from natural background radiation and wanted medical radiation to be exempt. Finally, in the late 1950s, the ICRP moved closer to the NCRP's position, recommending that the average dose to the public from all radiation, except background radiation and medical radiation, must not exceed 170 millirems per year. On the assumption that not everyone in the population will receive his or her full 'allotment' of non-background radiation, both the ICRP and the NCRP agreed to allow people who *are* exposed to receive up to 500 millirems per year (one-tenth the occupational limit). Thus was created a second double standard, one which treats members of the public differently according to whether or not they happen to live, play, or work near a nuclear facility.

An average exposure of 170 millirems per year will result in some genetic damage to the population, the ICRP acknowledged, but this could 'be regarded as tolerable and justifiable in view of the benefit that may be expected to accrue from the expansion of the practical applications of atomic energy . . . The Commission believes that this level [170 millirems per year] provides reasonable latitude for the expansion of atomic energy programmes in the foreseeable future.' But the ICRP issued a low-key caveat: 'It should be emphasized that the limit may not in fact represent a proper balance between possible harm and probable benefit', because of the uncertainty in assessing the risks and benefits that would justify the exposure. This was apparently a warning that, while the limit was fair for the nuclear industry, it might later be found to be unfair to the public.

CHAPTER FOURTEEN

Medicine After the Bomb

Since the 1940s, the technical focus of radiation had shifted from medicine to the bomb, but the technology that made the bomb possible also allowed a great expansion of the medical uses of radiation in the post-war years. One important advance was the development of the so-called supervoltage radiation therapy units capable of emitting very high-energy radiation. Such radiation is extremely penetrating, more of the dose is delivered where it is needed – deep within the body rather than at the surface. Even large doses do not burn the patient's skin. One unforeseen consequence of this 'skin-sparing' effect was that doctors had no visual signs that exposures might be dangerously high. A few doctors worried aloud that supervoltage therapy was being widely embraced before its long-term effects were known. In 1934, Dr Roswell Pettit warned his colleagues 'against too hasty conclusions drawn from insufficient experimental and clinical evidence', but his words and those of a few other sceptics were swept aside by enthusiasm for the possibilities of deep radiation therapy.

In 1948 James English was a young radiologist embarking on a career in San Francisco. He had just finished his residency at Stanford under the pioneer radiotherapist Robert Reid Newell, a founder-member of the postwar ICRP, and was bringing the latest ideas and techniques into his work. Supervoltage X-ray machines were part of his medical arsenal. Almost 40 years later, English sat in his abandoned radiotherapy room, its lead-lined walls now covered by wallpaper, and reminisced about the changes in his profession. 'When I started out, therapy was a big part of our work,' he told me. 'We treated malignant and benign conditions, but mostly cancer. For carcinoma of the larynx the normal course of treatment was every day, except Sundays, for six weeks. We aimed for a top dose of 4,500–5,000 roentgens [rems, in modern parlance] for the malignant diseases. For benign diseases we kept it to 400–600.' These were massive doses, though delivered only to part of the body.

Among the benign diseases then being treated with radiation were ringworm, acne, birthmarks, bursitis, and sinusitis. Between 1949 and 1960, doctors irradiated the scalps of more than 10,000 children in Israel, and a similar number in New York, who were infected with ringworm. The idea was to irradiate the child's scalp until the hair fell out, so that the ringworm could be treated more efficiently. 'That just scared the heck out of me,' said English, who refused to perform the operation; 'I drew the line there.' The average dose to each child's brain was approximately 150 rems, to the salivary gland about 40 rems, and to the thyroid less than 10 rems. The groups attracted researchers because they were large enough to yield statistically significant results and there were good treatment records and plenty of non-irradiated children for comparison. 'When they found that there were these large groups, with controls, the epidemiologists were eager to get in there and look at the numbers,' remarked Don Thompson, a medical physicist at the US government's Bureau of Radiological Health. What the epidemiologists found when they got in there was that the irradiated children had six times as many cancers of the thyroid as the controls did, as well as an excess of brain cancers and leukaemias. These studies, which were reported in medical journals, were among the first to alert doctors to the potential hazards of modern radiotherapy techniques, but they were followed by others with equally disturbing findings.

One of the most influential studies was a British survey of more than 14,000 people who were subjected to X-rays during the 1930s, 1940s and 1950s for a painful, but not fatal, back condition known as ankylosing spondylitis. Researchers found that 60 later died of leukaemia, though only 5.4 leukaemia deaths would be expected in such a population. The group also had a lung cancer rate twice as high as expected, and a significant excess of stomach cancer. Women who were castrated by radiation as a treatment for excessive menstrual bleeding and other benign gynaecological disorders were the subject of five studies. All five found that more of the women had died of leukaemia than would be expected and that there was an unusually high incidence of cancer of the intestines and of other organs that had been inadvertently irradiated during the treatment. Another study – of people whose thyroids had been treated with X-rays 30 or more years earlier – found that up to one-third of the chromosomes in their thyroid tissue cells were damaged.

As the implications of these studies sank in, said English, 'people began to doubt whether we should be using this kind of therapy for benign lesions. There was no question that [the treatment] was very useful. The question was whether there was something else that could produce the same effect and less injury. I was taught by Newell, who used to say, "It's not fair to make the patients sicker than they are to begin with."' English and his partners considered treating only malignant diseases, but eventually decided to give up X-ray therapy altogether and concentrate on diagnosis.

Even after specialists such as English began to refuse to treat benign ailments with radiation, thousands of doctors with little or no training in radiotherapy continued to use it. 'The overwhelming majority of people who were carrying out radiation therapy were not trained specialists,' says Dr Juan del Regato, who founded the first centre for radiotherapy training, at the Penrose Cancer Hospital in Colorado Springs. A survey made in 1973 found 44 per cent of American dermatologists still using X-ray therapy at least once a week for a variety of conditions, including squamous cell carcinoma (a mild type of skin cancer), lymphatic cancer, acne, eczema, psoriasis, and disfiguring scar tissue.

In 1912, the physicist George Hevesy was living in a boarding house in Manchester, England, while he worked at Ernest Rutherford's laboratory. The food was not very good – and, Hevesy was certain, not very fresh. To test his suspicions, one Sunday night Hevesy slipped a little radioactive lead from the lab into the leavings on his plate. Using a simple radiation detection instrument, Hevesy found traces of radioactivity in the meals served every night the following week. Hevesy's prank was an early experiment in the use of radioisotopes as 'tracers' – a technique that allows scientists to study chemical processes that are otherwise invisible. The technique, refined under laboratory conditions, won him the 1943 Nobel Prize.

Doctors were quick to see the possibilities of tracing the course of a radioactive substance through the human body. Unfortunately, nature did not make radioactive isotopes of such biologically important elements as carbon, phosphorus, calcium, and sodium, so scientists were unable to study the behaviour of these essential elements inside the body. But in the mid-1930s, soon after the news of Irène and Frédéric Joliot-Curie's discovery of artificial radioactivity, Ernest Lawrence at the University of California began using his invention, the cyclotron, to produce large quantities of artificial radioisotopes.

As medical agents, man-made radioisotopes have several advantages over radium, whose damaging side-effects were becoming increasingly apparent throughout the 1930s and 1940s. Radium–226 has a half-life of 1,600 years and is not easily expelled from the body, but researchers were turning up man-made isotopes whose half-lives and duration in the body were much shorter, making over-exposure of the patient less likely. Moreover, by choosing isotopes with the correct chemical composition, doctors could 'target' the organ that needed treatment.

'In those days we were able to find a new radioisotope every month or two,' said Glenn Seaborg, who then worked under Lawrence and later became the head of the Atomic Energy Commission. Lawrence and his team were even able to create made-to-order isotopes, as Seaborg recalled in the following anecdote:

> Dr. Joseph G. Hamilton . . . mentioned to me the difficulties he was having with radioactive iodine tracer in his experiments. At that time he was using the iodine–128 isotope, which has a half-life of only 25 minutes. He inquired as to the possibility of finding another iodine isotope with a longer half-life, which led me to ask him what value would be best for his work. He replied, 'Oh, about a week.' You may recall that soon after that we synthesized iodine–131, with a half-life of eight days.

Nuclear medicine did not really blossom until after the Second World War, when radioisotopes from the first nuclear reactors were made available to doctors at lower cost and in greater quantities than before. By 1952, the AEC had licensed more than 1,200 medical institutions to use reactor-produced radioisotopes. By 1960, radium was *passé* as a medical tool.

The most commonly used radioisotope was, and still is, iodine–131, which is mainly used to diagnose and treat cancer of the thyroid and hyperthyroidism, a non-fatal condition. Because the thyroid naturally absorbs iodine, it also takes up radioactive iodine, sufficient quantities of which will kill the cancerous or over-active thyroid cells. The alternative treatment for both conditions is surgical removal of the thyroid, but for some hyperthyroid sufferers, surgery can be dangerous or even fatal.

Since 1946 tens of thousands of people with over-active thyroids have been treated with I–131, receiving doses of up to 10,000 rads. But there is still some question about the side-effects of such treatment. Though most of the radioactive iodine goes to the thyroid, other organs are also exposed, so the treatment carries a certain risk of cancer. A study, published in 1974, compared the health records of 21,000 people treated with I–131 and 11,000 whose thyroids were surgically removed. The study concluded that there is no serious risk of cancer with I–131 therapy. However, the study was closed before the full effects of I–131 therapy could reveal themselves. Patients were followed for an average of only eight years after treatment, though cancer takes much longer than that to develop. This and other shortcomings in the study led the National Academy of Sciences to conclude that there is 'a serious question as to the validity of the conclusion that I–131 therapy is relatively harmless'. The population of Rongelap, the Marshall Island contaminated by I–131 fallout from the Bravo explosion in 1954, has been studied for a longer period. Detailed health records kept by the US government show that by 1980, one-third of the exposed population had developed thyroid lesions, even though the maximum radiation dose to the thyroid among the Rongelapese was less than 1,200 rads.

Compared to the radiation doses used in therapy, diagnostic exposures in the decades after the war seemed very small indeed. In theory, of course, no radiation exposure was totally safe, but in practical terms diagnostic X-rays were thought to be nothing to worry about. This relaxed attitude, combined with dangerous procedures and equipment, resulted in many people receiving unjustifiably large doses. Dr Francis Curry was deputy director and later director of public health and Hospitals in San Francisco from 1960 to 1976. 'What was so horrible about what was happening then,' he recalls, 'is that so many machines had no filters, no coning, no shielding. Many people were getting total body exposures and were getting doses big enough to show clinical symptoms.'

One of the worst abuses of X-rays was the rage for fluoroscoping children's feet in shoe stores. 'Little boys were having their gonads X-rayed,' says Curry. 'I remember one store where the manager was refusing to remove the fluoroscope. I put a dosimeter in my pocket while I was talking to him. A family of kids next to us was getting fitted out. Afterwards I told him, "I got more gonadal irradiation in the few minutes we've been talking than I have in a year of working with

X-rays." There was something else that amazed me. Someone suggested I look at the dog and cat hospitals. Veterinarians were building their own fluoroscopes, and making the owners hold the animals up right in front of the machines. These people knew absolutely nothing about radiation or X-rays. The exposures were unbelievable.'

In 1950, San Francisco, like many other cities across the country, established a mass chest X-ray programme intended to detect tuberculosis. The survey used photofluorograms – photographs of fluoroscope images – rather than conventional X-rays. If a photofluorogram indicated there might be a problem, the subject was then advised to have a more detailed X-ray examination. More than 40,000 people a year visited the city's clinics or a specially equipped van that toured the city, offering free chest X-rays. 'One reason for the large neighborhood programmes,' explained Curry, 'was that the lung associations wanted this type of continuing service. It would help them with their fund-raising. Local groups would have contests to see who had the most members screened. Lots of the black churches did this. Everybody and his brother would go. The same people went over and over again. They were getting a lot of radiation and we were getting no disease at all.'

Karl Morgan's group at Oak Ridge National Laboratory monitored some of the X-ray devices used in the mass screening and found that they were delivering skin doses of between 2,000 and 3,000 millirads, while the average dose from a chest X-ray unit at Oak Ridge was 15 millirads. There were other problems with the screening programme as well. Ninety-five per cent of the tuberculosis cases found had been detected without X-ray screening. Even when performed every six months chest X-rays are not sensitive enough to detect lung cancer before its symptoms appear. In 1965 the US Surgeon-General called for an end to such mass screenings. But it was declining productivity rather than the Surgeon-General's statement that finally caused the programme's demise, according to Robin Jones of the San Francisco Lung Association. 'We were taking 40,000 chest X-rays and finding one case of TB,' said Jones. By 1970 the mass screening programmes in US cities were largely over.

During the 1950s, some scientists became concerned about the growth of radiation as a medical tool. Geneticists who had warned of the dangers of fallout began to say that medical radiation was becoming an even greater hazard due to the abandon with which it was being used.

In 1956, the eminent geneticist Curt Stern warned that 'efforts should be increased to lower the average medical exposure'. ICRP member Hermann Muller, writing in *Science* magazine, said, 'Our people are annually receiving much more radiation [from medicine] than they do as a result of nuclear test explosions. A significant fraction of this radiation reaches the reproductive organs. Unfortunately, the majority of physicians have for 28 years closed their eyes to the genetic damage.'

Karl Morgan was an outspoken advocate for reform of medical radiation procedures and equipment during the 1960s. Morgan also wanted the ICRP to limit X-rays of women of reproductive age. This idea met with almost universal opposition from the medical community which held as sacred every doctor's right to treat patients without interference. It was for the doctor, doctors argued, to weigh the risks and benefits of using radiation, according to each patient's circumstances. In general the ICRP agreed with this view, but in 1962 Morgan and Muller persuaded the commission to recommend that, insofar as possible, radiological exposures of women should only take place within ten days of the start of menstruation, since that is the time when a women is least likely to be pregnant. The so-called 'ten-day rule' was – and still is – controversial in the medical profession. Britain's Royal College of Radiology adopted the ten-day rule, but the American College of Radiology never has. Several ICRP members spoke out against it on the grounds that it was impractical and of questionable medical value – and in 1983, the commission rescinded the 'ten-day rule' saying 'there need be no special limitation on exposures required within (the first) four weeks of pregnancy.'

The fallout debate caused many patients who had previously assumed that any medical radiation was safe to question their doctors' judgements about the need for radiation diagnosis or therapy. 'For a long time, people never questioned is this harmful, or could this be harmful?' recalled James English. 'Then people began to become very very frightened of X-rays. I think that part of it arose with the atomic bomb and people became more and more aware that this kind of radiation could be harmful. Numerous articles appeared in the lay press. Some patient was always coming in with an article from *Reader's Digest*, saying "Hey doctor, how about this?"'

Many radiologists resented this change in the public's attitude. They felt that fear of fallout and nuclear testing was unfairly being diverted to their work, largely by inaccurate and inflammatory journalism. And they feared that, left unchecked, the public's suspicion of radiation might hamper the use and development of one of medicine's most valuable tools. Already, there had been calls for an outright ban on the use of X-rays. Francis Curry remembers attending a lunch at San Francisco's all-male Bohemian Club, a meeting of radiologists worried about the public's mistrust of medical radiation. 'Most of them were saying the newspapers are to blame. I said, "No, they're right. Some of these exposures aren't warranted. We've got to start policing ourselves."'

CHAPTER FIFTEEN

The Peaceful Atom

In the public mind, atomic power has been linked with death ever since the destruction of Hiroshima. In the mid-1950s, when pressure was mounting in the United States and worldwide for international control of all nuclear materials, the US government resolved to improve the atom's public image. By promoting civil nuclear power and other non-military uses of the atom, President Eisenhower and his advisors hoped the public would associate the atom with peace and productivity, rather than with death, and would focus on the benefits of atomic energy, rather than the problems of weapons testing and radiation. Another important consideration in the push for nuclear power was the military's need for more plutonium than its own reactors could produce.

In a speech to the United Nations in late 1953, Eisenhower announced that the United States would share its nuclear expertise with the other nations of the world. Out of this sharing would come, not bombs, but efforts 'to apply atomic energy to the needs of agriculture, medicine, and other peaceful activities. A special purpose would be to provide abundant electrical energy in the power-starved areas of the world'. Eisenhower proposed that the United Nations establish an International Atomic Energy Authority to hold stocks of fissionable material – donated by the US, the Soviet Union and Great Britain – and to 'devise methods whereby this fissionable material would be allocated to serve the peaceful pursuits of mankind', so that 'the miraculous inventiveness of man shall not be dedicated to his death, but consecrated to his life'.

Eisenhower's eloquent proposal was generally enthusiastically received. In the media there was a burst of enthusiasm for the peaceful atom, as there had been twice before – in the early days of radioactivity and again at the end of the Second World War. News articles with headlines like 'Forestry Expert Predicts Atomic Rays Will Cut Lumber Instead of Saws' and 'Atomic Locomotive Designed' gave the bomb-

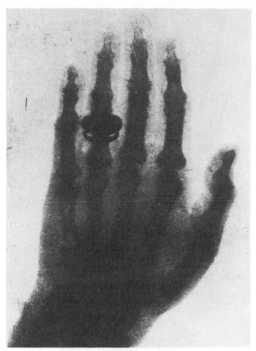

1. The first X-ray image: Frau Roentgen's hand
showing her wedding ring

2. X-ray exhibition
at Crystal Palace, 1896

3. Edison viewing a hand
through a fluoroscope

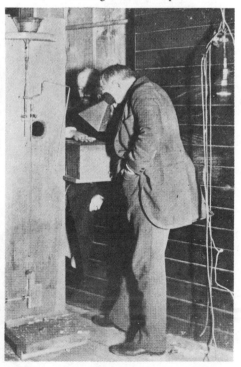

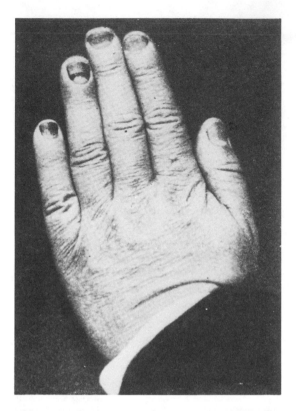

4. Doctor's hand showing
effects of continual exposure
to X-rays in testing equipment

5. Early X-ray equipment,
showing doctors studying
image on screen

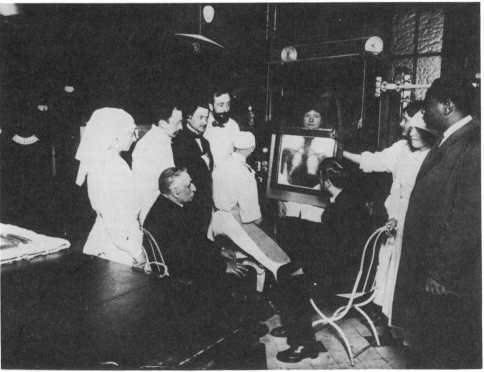

6. New outfit for X-ray work,
March 1909

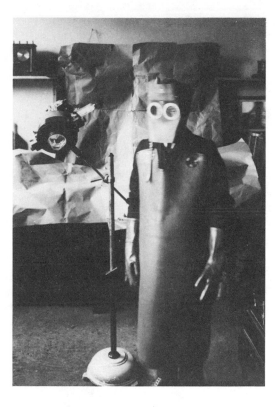

7. Weird radium treatment
mask

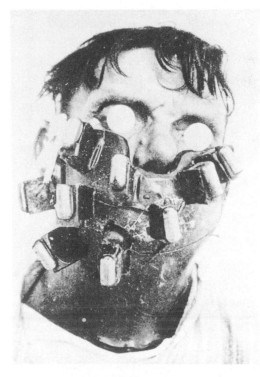

8. Advertisement for radium water apparatus

RADIUM THERAPY

The only scientific apparatus for the preparation of radio-active water in the hospital or in the patient's own home.

This apparatus gives a <u>high</u> and <u>measured</u> dosage of radio-active drinking water for the treatment of gout, rheumatism, arthritis, neuralgia, sciatica, tabes dorsalis, catarrh of the antrum and frontal sinus, arterio-sclerosis, diabetes and glycosuria, and nephritis, as described in Dr. Saubermann's lecture before the Roentgen Society, printed in this number of the "Archives."

Description.

The perforated earthenware "activator" in the glass jar contains an insoluble preparation impregnated with radium. It continuously emits radium emanation at a fixed rate, and keeps the water in the jar always charged to a fixed and measureable strength, from 5,000 to 10,000 Maché units per litre per diem.

SUPPLIED BY

RADIUM LIMITED,
93, MORTIMER STREET, LONDON, W.
Telephone: 4174 MAYFAIR.

instant X-ray sight

X-ray 'spec's that give you the amazing illusion to see right through everything you look at! See the bones in your hands, the yolk in an egg, the lead in a pencil and . . . . the most amazing things, when looking at girls and friends! Especially amusing at those fun parties!

£1.75 incl. P.&P.

Order soon so as not to be disappointed C.W.O. to:
SENE PARK PRODUCTS (SM7)
LAMBERTON HOUSE, SENE PARK,
HYTHE, KENT CT21 5XB.

COPYRIGHT TRACOMIN

9. Advertisement for X-ray spec's

10. Operation Crossroads: Test Baker, 25 July, 1946 (courtesy of The Bettman Archive)

11. Marines take part in atomic tests on Nevada Proving Ground, 1952

12. Nuclear fashion parade: a family models anti-radiation suits
in Wellingore, Lincolnshire

13. Amsterdam: a freighter loading vats of radioactive waste
to be dumped in the Atlantic Ocean

14. 'oh God, it rains!' reads the graffiti in Eppertshausen, West Germany, after ground radioactivity levels from Chernobyl fallout rose dramatically following rainfall

15. Contamination warning issued by health authorities in Germany following Chernobyl disaster goes unheeded

weary public a more optimistic view of how science would shape the future. Government-commissioned films with titles like 'Atomic Zoo', 'Atomic Greenhouse', and 'Atom for the Doctor', had the same effect. The General Electric Company, soon to be, with Westinghouse, one of the two major builders of nuclear power plants, distributed two million copies of a booklet called 'Atoms At Work' featuring 'your friend, Citizen Atom'. The Atomic Energy Commission even drew in the Boy Scouts, helping them to establish an Atomic Energy Merit Badge, which were awarded to tens of thousands of boys.

Many writers and commentators pushed aside questions about fallout to paint utopian pictures of a peaceful and productive atomic future. Glenn Seaborg, the discoverer of plutonium and later chairman of the AEC, was one of the most energetic promoters of the peaceful atom. Among his predictions: 'small atomic batteries' to power artificial hearts and other organs; nuclear-powered rockets to explore the moon; atomic slaves to 'do the dirty work' in industry and at home; 'synthetic food' produced without photosynthesis; abundant clean water from nuclear-powered desalinization plants. Lewis Strauss, then chairman of the AEC, presented his vision of the atomic future to a group of science writers in 1954. 'Our children will enjoy electrical energy too cheap to meter; will know of great periodic regional famines only as matters of history; will travel effortlessly over the seas and under them and through the air.'

One offshoot of the Atoms for Peace programme in the United States was the Plowshare Project, a sort of 'bombs for peace' plan. Plowshare was born during the Suez crisis of 1956–7, when the Egyptian government seized the Suez Canal. In a secret meeting at the Lawrence Radiation Laboratory in February 1957, a group of physicists, including Edward Teller, Harold Brown, later Secretary of the Air Force, and Ernest Lawrence, in whose honour the laboratory is named, decided to promote the use of nuclear explosives to cut a new canal across the Sinai Peninsula. The end of the Suez Crisis a few months later removed the need for a second canal, but it did not dampen the enthusiasm for using nuclear explosives as a civil engineering tool.

The AEC publicly launched Plowshare in 1958. Among the schemes proposed were the excavation of a harbour near Point Hope, Alaska; the blasting of a road and railway tunnel through the Bristol Mountains in California, the digging of a canal to link two separate river systems in Alabama and Mississippi, and – most ambitious of all –

the creation of a new canal across the isthmus of Panama, dubbed the Panatomic Canal. All these plans were eventually shelved for economic, environmental, and political reasons, but some smaller-scale 'peaceful explosions' were carried out. With the banning of atmospheric explosions, Plowshare enthusiasts focused on underground projects, primarily releasing deep deposits of natural gas. In 1967, a 26-kiloton demonstration blast called Gasbuggy was carried out in New Mexico. The experiment was a success in producing gas, but the gas was considered too radioactive to be used in homes or workplaces. Instead it was burned off at the well, and the radioactivity released to the open air. The Plowshare Project faded away when the AEC was abolished in 1975.

The main focus of Atoms for Peace was the production of electricity. The United States was determined to take the lead in the race for nuclear power, so that its designs and equipment would dominate the infant world market for nuclear power plants. Shortly after Eisenhower's speech, the 1946 Atomic Energy Act was amended by Congress to encourage commercial development of nuclear power. Within a week of the new act's passage, President Eisenhower inaugurated the construction of the nation's first commercial nuclear power plant, in Shippingport, Pennsylvania.

Both the Soviet Union and Great Britain had beaten the US to the punch with nuclear power. In June 1954 the USSR inaugurated the world's first nuclear power station, though its output was only 5 megawatts. The world's first full-size nuclear power plant was at Windscale on Britain's northwestern coast, bordered by peaceful dairy farms. Queen Elizabeth II formally opened the power plant on 17 October 1956. 'A-Power: It's Here! In Action! The First Dinners are Cooked!' trumpeted the *Daily Express*. One year later, the fuel in the core of a military reactor on the same site caught fire. The fire burned for 42 hours before anyone realized something was wrong; another 24 hours passed before it was extinguished. In the meantime, clouds of radioactive materials poured from the reactor's chimneystack. The fallout blanketed the surrounding countryside, and travelled as far as Denmark, Belgium, the Netherlands, France and Germany. The Windscale fire released more than 20,000 curies of radioactivity, thousands of times more than US officials say was released by the Three Mile Island accident in 1979. The main hazard was contamination by radioactive iodine of milk from cows grazing in the vicinity. The government banned the drinking or selling of milk from a 200-square-mile area for 25 days. Some cattle were slaughtered and half a million gallons of milk were dumped into local rivers. Britain's National Radiological Protection Board estimates that by 1997 the Windscale fire will have caused about 33 deaths, and 260 cases of thyroid cancer. Coming at the height of the weapons testing, the Windscale fire and its aftermath – particularly the ban on drinking local milk – drew attention to the dangers of contamination of the food chain by fallout, whether from bombs or from nuclear power reactors. Since the power industry had barely got under way, the public was more concerned about test fallout than about radiation from nuclear power plants. Nonetheless, from the very beginning the AEC faced serious opposition to its plans for nuclear power.

In 1956 the United Auto Workers tried unsuccessfully to halt construction of the Fermi reactor, located near Detroit, after a meltdown had destroyed the core of a research reactor at the same site the previous year. The case went all the way to the Supreme Court, which in 1961 affirmed the AEC's right to license the plant. Nevertheless, problems plagued the building of the Fermi reactor and on 5 October 1966, as engineers were attempting to bring it to full power, something went wrong. The exact nature of the mishap has never been made clear: nuclear critics say it was a meltdown that came within minutes of destroying Detroit; the nuclear industry says it was a minor incident that proved the effectiveness of the existing safeguards. In any case, the reactor was closed for repair for four years. The Fermi reactor was finally restarted in late 1970, but problems continued, and in 1972 it was permanently shut down.

The development of nuclear power in the US proceeded slowly at first. The Shippingport plant went into operation at the end of 1957, producing 60 megawatts of electricity. In the next five years, only three more nuclear power plants were built – one each in Illinois, Massachusetts, and New York. By 1970, nuclear power still provided less of the nation's energy than did firewood. That situation was due to change drastically during the 1970s, as the scores of large reactors authorized during the 1960s began operating.

One of the first successful protests in the United States against a nuclear reactor was not inspired by health or safety concerns. In 1962 conservationists in California organized to oppose plans to build a nuclear reactor at Bodega Bay, a beautiful stretch of coastline 50 miles north of San Francisco. 'We didn't know much about radiation at the beginning; our concern was the scenery,' recalled David Pesonen, one of the protest leaders. But the discovery that the site was only 1,000 feet from the San Andreas Fault led the protesters to look into the possible effects of an accident. 'We knew there would be an accident, so we studied the effects of the Windscale fire on the environment. As the campaign continued, radiation emerged as a critical issue,' said Pesonen. On May 30 1963 the objectors held a jazz concert at Bodega Bay and released 1,000 helium-filled balloons. To each balloon a postcard was attached that said, 'This balloon represents a particle of strontium–90 or iodine–131 or caesium–137 released from Bodega Head. When you find it please mail it to your nearest newspaper editor.' The balloons were found 'all over the place', according to Pesonen. Some came down at the Marin Civic Center, a Frank Lloyd Wright building across the bay from San Francisco; others wafted across the Mill Valley Country Club; a few floated into hotel rooms in San Francisco. In 1964, the power company, Pacific Gas & Electric, dropped its plans to build a reactor in Bodega Bay.

Most of the early reactors generated only about 60 megawatts (MW) of electricity, but within just a few years the AEC was issuing construction permits for reactors capable of generating 1,000 MW or more. As the reactors grew, so did the magnitude of the damage an accident could cause. An AEC study in 1956 concluded that for each reactor, every year, there was a 1-in-100,000 chance of a disastrous nuclear accident occurring. It appeared that these findings could end the development of commercial nuclear power, since the cost of insurance against damages on that scale would make nuclear power uneconomic. Congress, however, came to the rescue with the Price-Anderson Act, which it adopted in 1957. The act bans citizens from suing for damages caused by a nuclear accident. Instead it ordered the Atomic Energy Commission and civil power companies to set up a $560 million fund from which victims of major nuclear accidents would be compensated. The AEC's own study warned that a single nuclear accident could kill 3,400 people, injure 43,000 more, and cause $7 billion worth of property damage. The sponsors of the Price-Anderson Act settled for less than one-tenth that amount, because, as one of them later explained, a larger sum would imply a larger risk, and they did not want to 'frighten the country and the Congress to death'.

CHAPTER SIXTEEN

Confusion and Acrimony

The atoms for peace programme had created a privately owned nuclear power industry, but its future – and that of the military nuclear programme – was threatened by the public's growing concern about radiation. Throughout the 1960s, the AEC was under attack by a growing battery of scientific critics presenting evidence which they claimed proved radiation was more hazardous than officials would admit.

One of the most devastating blows to the commission was a study published in the *British Medical Journal* in June 1958 by Dr Alice Stewart, then head of Oxford University's Department of Preventive Medicine. Stewart became aware in 1955 that the number of children dying of leukaemia in England and Wales had risen by 50 per cent in the past few years, and that the increase was even greater in the United States. With her colleagues, Stewart interviewed mothers of 1,400 children who had died of cancer in England and Wales between 1953 and 1955. They also interviewed the same number of mothers with healthy children. Though not intentionally focused on radiation, the study turned out to be the first large-scale investigation of humans exposed to low doses of radiation. Stewart found that babies whose mothers had had pelvic X-rays while they were pregnant were almost twice as likely to develop leukaemia or another form of cancer as those whose mothers had not been X-rayed.

At a time when many scientists and physicians assumed that doses of 50 or even 100 rems were harmless, Stewart's study found that doses of the order of 1 rem significantly increase the likelihood of an unborn child developing cancer. Stewart's work thus seemed to disprove the AEC's theory that there is a threshold below which radiation is harmless, or at least to set the threshold impractically low. Her paper was widely and bitterly criticized. Detractors said that she had improperly relied upon the memories of the mothers she interviewed for important details about the number and timing of X-rays. 'The antagonism was from the medical profession,' Stewart recently recalled. 'Obstetricians and radiologists didn't like being told what they could do by me or by anyone else.' But in 1962 Stewart's work was confirmed beyond a doubt by Dr Brian MacMahon of Harvard's School of Public Health. MacMahon did not rely on the memories of mothers, but used hospital records of X-rays instead. His conclusions supported hers: children whose mothers had been X-rayed while pregnant were more likely to develop cancer.

Stewart's findings were ominous for the thousands of pregnant women and children who had already been exposed to supposedly safe levels of radiation through fallout. Exposure to X-rays could not account for the total increase in childhood cancers observed in Great Britain and the United States: might the remainder be attributed to fallout? To counter such speculation, the AEC decided in 1962 to establish its own programme to study 'man-made environmental radioactivity and the effects upon plants, animals, and human beings'. The study was to be conducted at the Lawrence Livermore Laboratory, which is run, under contract to the AEC, by the University of California. John Gofman, a physician and nuclear chemist, was chosen to head the new programme and was made an associate director of the laboratory.

Gofman had close ties to AEC chairman Glenn Seaborg. In the early 1940s, Seaborg was Gofman's thesis adviser at the University of California at Berkeley. In 1942 Seaborg, Gofman, and Raymond Stoughton discovered four important radioactive isotopes, including uranium−233, a valuable nuclear fuel. Gofman, who had developed a way of separating plutonium cleanly from uranium, gave Robert Oppenheimer his first milligram of plutonium at a time when the entire world supply of plutonium was about 0.06 milligram. Since leaving the Manhattan Project, Gofman had concentrated on problems of heart disease.

At Lawrence Livermore, Gofman and his team pursued a dual course, studying ionizing radiation's effect upon chromosomes, and analysing data in the scientific literature on the effects of radiation on humans. Eventually they became convinced that the risk of cancer from radiation exposure was, in Gofman's words, '20 times greater than was being said in official circles'. They had not yet published their findings when Gofman's associate, Dr Arthur Tamplin, ran afoul of the management of the Livermore Laboratory and its chief funder, the AEC.

The dispute arose, ironically, as a result of Tamplin's efforts to refute one of the AEC's most strident critics, Dr Ernest Sternglass, a physicist at the University of Pittsburgh. In the fall of 1968, Sternglass submitted a paper to *Science* in which he calculated that between 1951 and 1966, fallout from testing had caused the deaths of 400,000 infants under the age of one in the United States alone. The paper was rejected by the then-editor of *Science*, Philip Abelson, a nuclear chemist who served on both the AEC's General Advisory Committee and its Project Plowshare Committee. Sternglass's paper was published instead in the April 1969 issue of the *Bulletin of Atomic Scientists*. A few months later, Harold Hayes, the editor of *Esquire* magazine, asked Sternglass to write an article on the same subject for his September issue. *Esquire* sent a copy of the issue to every US congressman and placed full-page advertisements in the *New York Times* and the *Washington Post* summarizing the main points of the article, which was titled 'The Death of All Children'.

Stung by Sternglass's widely publicized criticisms, AEC officials sent his paper around its labs, asking for comments. Arthur Tamplin, a biophysicist working with Gofman, reviewed the study and concluded that Sternglass had overstated his case. Sternglass had based his conclusions on the fact that in recent decades infant and foetal deaths in the United States had been decreasing at a constant rate, due to improvements in health care. During the period of testing, however, infant deaths, though continuing to decline, did so more slowly. After the test ban treaty, the rate of decline rose again. Sternglass attributed the 'extra' 400,000 infant deaths during the atmospheric test era to fallout. Tamplin found that Sternglass had ignored important social and economic factors affecting the infant mortality rate. Fallout killed perhaps 4,000 American infants, said Tamplin, not 400,000.

AEC officials welcomed the refutation of Sternglass, but not the conclusion by an AEC researcher that fallout had killed thousands of American infants. John Totter, who headed the AEC's Division of Biology and Medicine, and his boss, Spofford English, contacted both Tamplin and *his* boss, Gofman. They argued that Tamplin's plan to publish his findings in the *Bulletin of Atomic Scientists*, a journal whose audience includes politicians and journalists as well as scientists, was inappropriate. The Sternglass critique, they suggested, was appropriate for a general interest magazine, but the new infant and foetal mortality estimates should be published separately – in a specialist journal. Tamplin disagreed, in his words, because 'Sternglass's somewhat exaggerated claims were being contradicted with even more exaggerated claims by the AEC'. Backed up by Gofman, who says he told Totter and English 'to go to hell, essentially', Tamplin published his paper in the *Bulletin of Atomic Scientists.*

Meanwhile, Gofman was scheduled to present a paper to a meeting of the Institute for Electrical and Electronic Engineers in San Francisco on 29 October 1969. He and Tamplin decided 'to use this talk as an occasion to present our views on what the evidence seemed to show'. They told the meeting that 'if the average exposure of the US population were to reach the allowable 170 millirems [annual limit], there would, in time, be an excess of 32,000 cases of fatal cancer plus leukemia per year, and this would occur year after year'. True, the actual exposure was then only about one per cent of that limit, but Gofman and Tamplin argued that the limit was likely to be approached as more and more nuclear power plants were built, and as the use of radiation in medicine and other industries grew. The biological effects of radiation were 20 times worse than was realized when the exposure guidelines were adopted, they said, therefore the permissible dose to the population should be cut by at least a factor of ten. 'To continue the present guidelines is absolute folly,' Gofman told the meeting. Soon afterwards, the two repeated their warning to a Senate subcommittee.

Gofman's and Tamplin's charges made radiation safety an issue not only for people exposed to fallout, but also for workers in commercial power plants and people who happened to live near nuclear facilities of any type. Moreover, their call for a tenfold reduction in the legal radiation limit was simple enough for journalists, politicians, and the general public to understand. The AEC had to respond, though it was in the awkward position of refuting its own experts. On 17 December 1969, the commission's staff issued a formal rebuttal. It said that Gofman and Tamplin should not have extrapolated the effects of low doses of radiation from data on the effects of high doses of radiation, and that 'the responsible radiation-protection bodies' disagreed with Gofman and Tamplin. The commission's report concluded that 'the opinions and scientifically questionable derivations of Gofman and Tamplin do not make a case for revision of radiation-protection standards'. Informally, AEC officials told journalists that the two men were 'incompetent'.

At this time, Bo Lindel, then vice-chairman, later chairman of the ICRP, wrote a private letter about the Gofman–Tamplin claims:

I am not in agreement with Gofman and Tamplin, neither when
they propose a nuclear power station moratorium nor when they
suggest that the maximum permissible dose [MPD] should be zero.
But I find nothing to laugh at in these proposals . . . Those who
have listened know that the Gofman–Tamplin proposal is meant
to be a radical shift from the present situation where everything is
permitted, as long as the MPD is respected, even if the use of
radiation is entirely unnecessary or against the interest of most
other people, to a new situation where nothing is permitted until
the prospective user of a radiation source has convinced society
that the benefits outweigh the risks . . . I have heard many proposals which have been more stupid and I am ashamed on behalf of
the established radiation club that we have not been willing to
engage in a constructive discussion. I think that the G–T proposal
is a beautiful thought but I also think that it is naive and premature
in the present world where nothing else is governed by such
philanthropic rules.

With controversy swirling around them, Gofman and Tamplin
resumed their work at Lawrence Livermore. Whenever possible they
also appeared at congressional hearings and public meetings to press
for a tenfold reduction in radiation exposure levels. Their relations
with the laboratory management and with AEC staff grew increasingly
bitter. What Gofman calls 'the beginning of the end' came towards the
close of 1969, when officials at the Livermore Lab tried to bar Tamplin
from a meeting of the American Association for the Advancement of
Science unless he agreed to make extensive changes in a paper that he
was to deliver on the subject of nuclear reactors and the public health.
A furious Gofman accused lab director Michael May of running 'a
scientific whorehouse' and threatened to make formal accusations of
censorship. The lab eventually backed down, and allowed Tamplin to
deliver his paper with few changes. Two weeks later Tamplin's staff of
eleven was cut to four. Within six months he had only one assistant, and
his major responsibilities were transferred to other scientists.

In an interview with *Science* at the height of the dispute, Gofman said
that the AEC had, by harassing him and his co-workers, 'partially
succeeded in silencing our presentation of the radiation hazards issue'.
The AEC denied the charges, but could not contain the controversy.
The *Washington Post* picked up the story and Gofman and Tamplin
became a *cause célèbre*. In 1973, after the AEC threatened to delete a
quarter of a million dollars from Lawrence Livermore's budget if

the laboratory continued to support Gofman's research, he returned to full-time teaching at the University of California at Berkeley, where he is now Professor Emeritus of Medicine. In 1975, Tamplin too resigned and went to work for the Natural Resources Defense Council, an environmental research and lobbying organization in Washington, DC.

Most of the human populations whose response to radiation has been studied have received doses of 100 rems or more. It is much harder to obtain statistically reliable information about lower doses – doses from long-range fallout or ordinary X-rays, or emissions from nuclear facilities. One problem is that there is little reliable dose data for exposures of less than 10 rems. Researchers need to know the size of the dose, the rate at which it was delivered, what part of the body was exposed, and the age and health of the individual during exposure, among other things. There are no dose records for most people who live near nuclear power stations, weapons factories, waste dumps, or the Nevada test site. And exposure records that do exist – for military participants in nuclear tests for example – are often incomplete or unreliable.

Radiation-induced cancers are indistinguishable from all other cancers; they can only be detected statistically, in relation to the number of cancers that would normally be expected in a given population. Industrial societies have high rates of cancer deaths: in Great Britain 200,000 people die of cancer each year; the United States has about 500,000 cancer deaths annually. With such a large base of cancers, it may be statistically impossible to detect a relatively small number of *extra* cancer deaths. Some scientists argue that in order to get statistically reliable data on the effects of doses on the order of one rem, it would be necessary to study ten million exposed people, and a matching group of ten million people who have not been exposed to radiation. Matters are further complicated by the need to continue a study from the first exposure until death. With low-level exposures, a cancer may not appear for 30 or more years after it was induced.

Because of these difficulties, most risk-estimates for very low-level exposures are extrapolated from data on exposures at higher levels. Projecting from the known to the unknown is a matter of judgement and interpretation, not simply of mathematics. The ICRP says it is safest to assume that for all types of ionizing radiation there is a linear relationship between dose and risk – that is, a 1-rem dose is one-tenth as dangerous as a 10-rem dose. The US National Academy of Sciences assumes that, for gamma and X-rays, low doses cause less damage, unit for unit, than high doses – that is, a 1-rem dose is *less* than one-tenth as risky as a 10-rem dose. Some of the data from the study of Japanese A-bomb survivors supports the idea that low doses of all ionizing radiations cause proportionately *more* damage than high doses – in other words that a 1-rem dose is more than one-tenth as dangerous as a 10-rem dose.

It also appears that cells are sometimes able to recover from low doses of radiation. One hotly contested point is what effect this ability has on the dose–response relationship. To some researchers, it clearly indicates that, unit for unit, low doses of radiation cause less damage than high ones. But others argue that a recovery mechanism makes low doses disproportionately *more* damaging than high doses, because, like a broken plate that has been glued back together, the repaired cells survive in a weakened condition. Though they may not cause cancer, the cells are more susceptible to infectious diseases and other assaults.

This notion that radiation exposure may be causing undetected injuries, such as a general lowering of resistance to disease, has not been much tested, partly because a generalized decrease in resistance to disease would be even harder than cancer to relate statistically to increased radiation exposure. Recently, however, Jay Gould, a respected American statistician, discovered a strong correlation between increased levels of radioactivity in some parts of the United States and higher-than-expected death rates. There were between 35,000 and 40,000 more deaths during the summer of 1986 than is usual for that time of year. Gould found that the areas with the greatest increase were those with the highest fallout from Chernobyl. One explanation for the correlation is that cells damaged by radiation lose their ability to produce white blood-cells, which are crucial to the body's immune system. This theory is supported by Gould's finding that the excess deaths involved a disproportionate number of old people and victims of infectious diseases.

In any study of the effects of radiation, there are many choices to be made that will influence the outcome of the study, sometimes in predictable ways. If a subject died of cancer, should the investigator disregard the last years of exposure on the grounds that the cancer was induced long before death, and that later exposures were superfluous? Such an approach would clearly link cancer induction with lower levels of exposure. Likewise, the choice of a 'control' population with which to compare the exposed population will influence the study's findings. It is important that the control group be as closely matched to the exposed group as possible. For example, many researchers believe that radiation workers are healthier than the general population, because they must pass rigorous physical exams to be hired. If the so-called 'healthy worker effect' is accepted, then the control group too must be healthier than the general population.

One of the largest studies of low-level exposures was begun in 1965, when the AEC asked Dr Thomas Mancuso, a pioneer in the field of studying long-term health effects, to study the effects of radiation exposure on workers at AEC installations, particularly the Hanford Nuclear Reservation in Washington State. The Hanford works include a reactor complex dedicated to the production of plutonium and a nuclear waste dump. The Health and Mortality Study, as Mancuso's project was called, was slow work because of the need to assemble and analyse millions of bits of hard-to-obtain data dating back to the mid-1940s. In any case, Mancuso wanted to carry the study into the 1970s, to give long-latency cancers time to appear. The AEC, however, was eager for Mancuso to publish. 'I have been repeatedly urged by the AEC . . . to publish the negative progress reports,' Mancuso told a congressional hearing in 1978. 'I refused to publish prematurely in view of the long latency periods of cancer . . . Any analysis which did not meet the number of years required to induce the occupational cancer would lead to false negative findings that would be misleading and could be misused.'

Confidential AEC memos later obtained under the Freedom of Information Act indicate that the agency had commissioned Mancuso's study for political rather than scientific reasons. Some scientists who reviewed the study for the commission thought that it would be impossible to detect the rare and long-latency period cancers involved in the time allotted. According to a 1967 AEC memo, the reviewers believed that 'aside from a certain "political" usefulness, it was very unlikely that new information on radiation effects will accrue from this study'. In the words of one reviewer, 'much of the motivation for starting this study arose from the "political" need for assurance that AEC employees are not suffering harmful effects'. AEC officials also hoped that the results of the study could be used to fight workmen's compensation claims for radiation-related injuries.

In the summer of 1974, Dr Samuel Milham of Washington State's

Department of Social and Health Services found an excess of cancers among Hanford workers, which he theorized may have been caused by their occupational exposure to radiation. Milham planned to publish his conclusions, but first he presented them to the AEC. Alarmed, the agency asked Mancuso to issue a press release to refute Milham. An AEC official dictated to Mancuso over the telephone the statement that they wished him to release. Mancuso was to say that Milham's conclusion 'is contrary to' his own finding that there is not an excess of 'cancer or other deaths attributable to ionizing radiation . . . among Hanford workers'. Insisting that he could not legitimately make such a statement until he had completed his analysis, Mancuso refused. The following March, Mancuso was informed that the AEC would not renew his contract when it expired in July 1977.

The AEC commissioned Battelle Northwest Laboratories, a research institute with close links to Hanford, to review Milham's data. According to a memo written by the manager of the Hanford Research Laboratory, the findings were unfavourable to the AEC: 'the message is clear that Battelle's data suggests that Hanford has a higher proportion of cancer deaths for those under 65 than the US'. Unfortunately, he wrote, 'we hoped to get a good answer to the Milham report, and instead it looks like we have confirmed it'. The Battelle report was not released.

In the last year of his contract, Mancuso recruited Alice Stewart and George Kneale, a statistician, from England to help him analyse his data on approximately 35,000 Hanford workers. Their analysis confirmed the findings of Milham and Battelle Northwest that, though most of the exposures were well within the legal limit, there were approximately six per cent more deaths from certain types of cancers among Hanford workers than would be expected. The MSK study, as it became known (for Mancuso, Stewart, and Kneale), linked cancers to radiation exposures previously widely assumed to be safe.

Now that Mancuso was ready to publish, the AEC didn't want him to. 'They urged us not to publish . . . My job in their eyes was simply to transfer the data to them,' he told an interviewer. Nevertheless Mancuso, Stewart and Kneale published their findings in the November 1977 issue of *Health Physics*. Two months earlier the AEC had terminated Mancuso's contract, and transferred the Health and Mortality study, without the usual peer review process, to two AEC-funded labs. Most of the research was assigned to Oak Ridge Associated Universities, a consortium formed to carry out research for the

AEC's Oak Ridge weapons laboratory, and Battelle Northwest. An AEC employee, Dr Sidney Marks, was given special exemption from federal conflict-of-interest statutes so that he could go to Battelle as the new project director for the study.

The Mancuso study spawned a mini-industry. Over the next several years, at least ten re-analyses of Mancuso's data were published, several commissioned by the Department of Energy, which took over responsibility for atomic weapons development in 1975 after the AEC was abolished. In addition, the original authors revised their analysis twice to answer criticisms of their methodology. Among the chief criticisms of the MSK study is that the authors used novel methods of analysing their data. For example, they compared the exposures of those Hanford workers who died of cancer with the exposures of those who died of other causes, rather than comparing the number of cancer deaths at Hanford with the number of cancer deaths in the general population, as is usual. Their critics say that this approach ignores the question of whether there was an excess of cancers in the first place. Mancuso and his co-authors maintain that their approach was necessary to adjust for the so-called healthy worker effect, the fact that radiation workers are an unusually healthy sub-set of the population. Critics also charge that the MSK study unfairly assumes that the fatal cancers were caused by radiation, rather than by something else, perhaps smoking or exposure to some chemical. The two sides have gradually moved closer to one another, and there now appears to be general agreement that while there is not a significant excess of overall cancers among Hanford workers compared to the general population, there is a significant excess of multiple myeloma and cancer of the pancreas, both of which can be induced by radiation.

Not surprisingly, with all the possible variables, the risk estimates made by different individuals and groups differ widely. The usual way to assess risk is by stating the number of deaths that would result from one million person-rems of exposure. (A person-rem is the unit for measuring collective dose. One thousand person-rems may be the collective dose of ten people whose average exposure is 100 rems, or of 10,000 people whose average exposure is one-tenth of a rem. Either way, the number of deaths and injuries in the exposed population should be the same.)

The United Nations Scientific Committee on the Effects of Atomic

Radiation (UNSCEAR) estimated in 1977 that a collective dose of one million person-rems of whole-body radiation will cause 100 deaths from cancer. The ICRP puts the figure at 125 deaths; the US National Academy of Sciences' committee on the Biological Effects of Ionizing Radiation (the BEIR committee) says 226 deaths; Karl Morgan says 1,000 deaths; John Gofman says 3,700 deaths. If non-fatal cancers, which are generally believed to be about one-and-a-half times as common as fatal cancers, were included, these risk estimates would range from a low of 250 to a high of 9,250 cancers per million person-rems of exposure.

The genetic risk of radiation exposure is even harder to pin down, because there is less data on genetic damage (damage that is transmitted to succeeding generations) than there is on cancer effects. No studies conclusively link radiation exposure to genetic damage in human beings. There are many studies showing that radiation causes genetic damage in animals, but it is difficult if not impossible to extrapolate these findings to humans. Genetic damage can manifest itself in a multitude of ways. About ten per cent of all babies are born with some damage to genes or chromosomes, but there is no way of knowing what proportion, if any, of this 'naturally occurring' genetic damage is due to radiation exposure. Evidence of genetic damage may not appear for several generations, so it is even harder than cancer to study. Many ailments, such as asthma or heart disease, are known to have a genetic component, but there are many others for which the genetic link is still uncertain.

Most authorities are too uncertain to even give risk estimates for genetic damage, but the ICRP estimates that one million person-rems will cause 40 genetic effects in the first few generations, and another 40 in all subsequent generations. The National Academy of Sciences' BEIR committee gives a range of figures; the average is that one million person-rems will cause 40 'serious genetic disorders' in each successive generation.

All these estimates were recently thrown into question when scientists at Lawrence Livermore Laboratory discovered a major error in the study of the Japanese A-bomb survivors, which is the main basis for most radiation risk estimates. The study, which until 1975 was funded by the US Atomic Energy Commission and overseen by the US National Academy of Sciences, is now run by the Japanese government's Radiation Effects Research Foundation (RERF), which receives half its funding from the United States government. The RERF has followed some 100,000 A-bomb survivors since 1950. In 1980, the Livermore scientists found that the estimates of the amount of radiation released at Hiroshima were wrong. The revised dosimetry indicates that gamma radiation is roughly 40 per cent more likely to cause cancer than previously believed.

In 1987, Dale Preston and Donald Pierce of the RERF published a re-evaluation of the 1980 BEIR risk estimates, using the revised Hiroshima dose calculations and new RERF data about cancers that had developed in people who were very young at the time of the bombing. Preston and Pierce used two different types of extrapolations on their data. Using the BEIR committee's extrapolation, which assumes that radiation is less damaging when delivered in small doses, they found that one million person-rems will cause 520 fatal cancers – more than four times the ICRP's risk estimate. Using a linear extrapolation, which assumes that a given amount of radiation causes the same amount of damage whether it is delivered in a few large or many small doses, the pair found that a million person-rems causes 1,300 fatal cancers. This figure is more than ten times higher than the ICRP's estimate. Edward Radford, who was until recently the RERF's chief of research, also gives an estimate of risk based on the latest RERF data that is ten times higher than the ICRP's estimate. Another RERF study, published in August 1988, finds that a million person-rems will cause 1,000 fatal cancers. That study also finds that a given amount of radiation is more damaging when delivered in small doses than in very high doses – in the words of the report, the data 'imply a higher risk at low doses'.

In the early 1970s, Arthur Tamplin wrote of 'an illusion that there is a radiation effects controversy'. In this he was supported by the ICRP's Bo Lindel, who wrote that Tamplin and Gofman's controversial risk estimates 'do not yield results which are very far from present risk numbers discussed within the commission'. With their bitter arguments, the two sides still give the impression that their scientific differences are irreconcilable, but the revisions to the RERF data are certainly bringing the radiation 'establishment' closer to its critics. Although John Gofman's risk estimate of 3,700 cancer deaths per million person-rems is 30 times greater than ICRP's, it is less than three times greater than Preston and Pierce's linear extrapolation figure of 1,300 cancer deaths.

The arguments about protection standards involve even smaller scientific differences. Most critics say that the ICRP standards are too high by a factor of from five to ten. But several members of the ICRP have already said that the present limits are likely to be at least halved, in view of the revisions to the A-bomb studies. According to Roger Berry, a British member of the commission, 'from several independent sources we are coming to relatively consistent estimates in the rate of cancer induction as a function of radiation dose. Now, whether those final figures are a factor of two, or a factor of five different from the current ICRP estimates remains to be determined.' In 1987 the NCRP cautioned that its own risk estimates 'will probably increase, perhaps substantially in the near future.'

The polarization of the debate about low-level radiation is due less to dramatically differing scientific views than to the practical implications of changing radiation protection policy. Indeed, the stakes in the debate are extraordinarily high. Billions of dollars and an unknown number of human lives, now and in future generations, rest upon the outcome. If it is discovered that a given amount of radiation is ten times more likely to cause cancer or 100 times more likely to cause genetic defects than officials have believed, then governments, as well as any company or hospital that makes or uses radiation equipment, would be more vulnerable to lawsuits by people seeking redress for radiation injuries. It would be even harder to find acceptable sites for nuclear reactors, nuclear waste dumps, or other facilities where radiation is used. Labour unions would certainly demand higher pay for jobs that increase the likelihood of a worker's developing cancer or his or her children being born with genetic defects. There would be pressure for better monitoring and record-keeping procedures, for a tightening of exposure limits, and for the remodelling of existing nuclear plants to enable them to meet the new standards – if not for the closure of all nuclear facilities and a dramatic shift in defence strategies.

The cost of these changes could make certain nuclear technologies uneconomic and result in the crash of one or more industries and the communities they support. They could create an extraordinary burden on taxpayers, energy-users, hospital patients, and ordinary consumers. Too conservative an approach to radiation protection would mean that these heavy costs were incurred pointlessly, while other, more urgent risks went unabated. If the critics are right, however, failing to reduce radiation exposure now will ensure that our generation and those that follow are burdened with the incalculable social and financial burden of radiation-induced diseases, a damaged genepool, and inescapable long-lived radioactive pollution.

CHAPTER SEVENTEEN

A House Divided

Every country with a nuclear industry has at least one nuclear regulatory agency; some have many more. In the United States, at least 16 federal agencies and 20 congressional committees, not to mention each of the 50 states, have responsibility for some aspect of radiation protection. The United Nations has at least four agencies concerned with radiation exposure – the Scientific Committee on the Effects of Atomic Radiation, the World Health Organization, the International Atomic Energy Agency, and the UN Environment Programme. Then there are the scientific and medical societies, the university labs and government-funded research institutes, and the industrial associations. Together, these groups make up what its supporters call the radiation protection community and its critics refer to as the nuclear establishment.

At the centre of the radiation protection community is the International Commission on Radiological Protection, a unique group with no legal power, but with almost unquestioned international authority. The commission is a private scientific body, established in 1928 to provide doctors and scientists with guidance on radiation safety. The ICRP's independence from governments and the prestige of its members make the ICRP attractive to nations in need of credible radiation protection standards. Its recommendations have been adopted with only minor modifications by virtually every country in the world (though the United States also heeds the advice of a similar and overlapping American group, the National Council on Radiation Protection).

Alone among radiation protection bodies, the ICRP makes political judgements, as well as technical ones – not only defining the risks of different radiation exposures, but also deciding how *much* risk is acceptable. Before the Second World War, commission members were accepting risks on behalf of themselves and their professional colleagues. But now, in recommending standards for blue-collar workers and the general public, the commission accepts risks on behalf of millions of people who don't even know it exists. (A few years ago there was a move to establish a unit for measuring risk based on a 1-in-10,000 chance of premature death. The unit was to be called a failla, in honour of the late Gioacchino Failla, a pioneer of radiation therapy and a long-serving member of both the ICRP and the NCRP. The idea was quietly dropped when friends of Failla protested.)

The commission does not aim for absolute safety: that would require banning all radiation exposures, a move which would conflict with the commission's philosophy that the benefits of nuclear technologies are too important to lose. Rather, its standards are intended to ensure that the risks of an exposure will be balanced by its benefits, though the risks and benefits are rarely evenly distributed among the population.

Robert Stone, the former head of the Manhattan Project's Health Division and an ICRP member, recognized the illogic of this situation in 1958, but concluded that there was no alternative. 'Maybe the facts should be laid out and many people express themselves on the risks that should be taken, i.e. on the maximum permissible dose,' he wrote, but 'I do not believe that such a course is possible, at the present time, because we still have to assess the reliability of the "facts" that come to us'. Even today, says Lauriston Taylor, the ICRP cannot simply compile a list of exposures with their attendant risks and let the public or their elected representatives decide what level of risk to accept, 'because we don't have the information. We can't tell you what's going to happen at five or ten rems a year. All we know is we cannot find any evidence of injury [at those levels]'.

The ICRP does not conduct research. It bases its recommendations on its evaluation of existing data and on the expertise of its members. Four 15-person standing committees, each headed by a member of the main commission, prepare the detailed technical reports which, after approval by the commission, are issued as 'Annals of the ICRP'. These reports tend to be favourably received by governments since all 13 members of the main commission, and many members of its four standing committees, are also national advisers on radiation safety. John Dunster, for example, was, until his recent retirement, both a member of ICRP and director of Britain's National Radiological Protection Board, charged by the British government with evaluating the ICRP's proposals. Dunster himself has remarked that 'there is a certain amount of incest in the business'.

The ICRP is a select body. Fewer than 50 scientists, all men, have served on it since the Second World War. More than half of them have been members for at least 10 years; five, including the current chairman, have spent at least 20 years on the commission. If time spent on the commission's sub-committees is included, then two-thirds of the commission's post-war members have served for 10 years or longer. Lauriston Taylor – now an emeritus member – has been on the commission since its founding in 1928. The ICRP is self-perpetuating. It appoints its own members, who are generally plucked from one of the commission's sub-committees. The International Congress of Radiology, a consortium of medical radiology societies under whose auspices the ICRP was established, has the right to veto new members, but it has never exercised this right. The present commission consists of three Americans, two Britons, and one representative each from Argentina, France, West Germany, China, Poland, Italy, Japan, and the Soviet Union.

A small, run-down wooden outbuilding on the grounds of the Royal Marsden Hospital in a south London suburb serves as the commission's headquarters. The staff of this powerful international body consists of the scientific secretary and a typist. Inter-governmental bodies like the International Atomic Energy Agency and the World Health Organization of the United Nations, as well as scientific organizations like the International Radiation Protection Association, give the commission about $150,000 per year. Employers of commission members, primarily government-funded agencies such as the Oak Ridge National Laboratory in the United States and the Medical Research Council in Great Britain, also help by underwriting research and providing travel expenses and administrative assistance. David Sowby, who retired in 1985 after 23 years as the commission's scientific secretary, estimates that the ICRP gets at least $1.5 million-worth of such aid each year.

The world's major source of funds for nuclear research is the US Department of Energy (DoE), successor to the disbanded Atomic Energy Commission. The DoE is responsible for the design, production and testing of nuclear weapons. It has nearly 40 research laboratories, employing more than 60,000 people. The annual budget for these laboratories is $10 billion. Many are operated by colleges or universities. The University of California, for example, receives more than $1 billion per year from the DoE to run two weapons laboratories, Lawrence Livermore in California and Los Alamos in New Mexico, and another $133 million for running Lawrence Berkeley Laboratory and three smaller nuclear labs. In 1985, the DoE spent $140 million on research into the health and environmental effects of radiation. 'There are very few people in radiation protection in this country who are not funded by the Department of Energy at one time or another,' says Richard Guimond, the director of the Environmental Protection Agency's Radiation Standards Division.

Radiation researchers who, like Thomas Mancuso and John Gofman, clash with the DoE are unlikely ever to get another federal research contract or work in an institution that depends upon federal funding. This intolerance of dissent has forced a fair number of experienced scientists into a sort of scientific exile, where they constitute a formidable opposition to the prevailing views. An article in the magazine *Science and Public Policy* names more than 30 researchers in ten countries, including India, Japan, Great Britain, the Soviet Union, and the United States, who were disciplined, fired, or lost their research grants after criticizing the nuclear establishment. Some critics find work in non-nuclear institutions, others are tenured university professors, quite a few are retired, but all find it difficult to get adequate research funds, and in some cases, access to government-controlled data.

'If you mind your own business and publish your research in obscure journals and keep quiet, you can get away with more,' says Dr Rosalie Bertell, a Canadian nun, statistician-biologist, and strong critic of the current radiation protection standards. Publicizing a disagreement or a divergent point of view is perhaps the most provocative thing a dissenter can do. Public quarrels weaken bodies like the NCRP and the ICRP, whose authority derives from their claim to represent a scientific consensus. These groups, in fact, tend to assume that dissent from the majority view is *ipso facto* evidence of scientific inadequacy. 'It's just an automatic thing,' observes Dr Robert Blackith, a quantitative biologist at Trinity College, Dublin. 'If you criticise the ICRP, they exclude you from the ranks of "serious" scientists.'

In an almost painfully thoughtful letter to his fellow ICRP members in 1971, Bo Lindel, then the commission's vice-chairman, discussed the ICRP's response to its critics.

> We react like mechanical puppets or like insects shown a stimulus triggering aggressive reactions, when we are faced with statements or ideas which are not branded with the mark of the old truth.

Should we not instead be curious and appreciative? . . . News media in the USA and throughout the world are full of records of a debate which was started by insignificant contributors but has been carried on persistently by individuals such as Sternglass, Gofman and Tamplin. In the ICRP records these persons hardly exist and we have reacted in a very highbrow and rather stupid way . . . Who are we that we can afford to smile at this from our lofty distance? . . . To the outsider it looks like a well protected mafia of big power interests, operating without being scrutinized and criticized.

As a founder and the longest-serving member of both the NCRP and the ICRP, Lauriston Taylor has become chief advocate of the radiation protection establishment and a fierce critic of its critics. He dismisses the dissenters as 'a small handful of pseudo-scientists who have been rejected by their past colleagues', 'a few hungry charlatans', and 'one-time scientists'. Taylor describes Joseph Rotblat, emeritus professor of physics at the University of London, as being 'part of this group of rather sloppy people that are using radiation to promote . . . a personal mission, an anti-war mission, heaven knows what'. Of Edward Radford, former chairman of the prestigious BEIR committee and professor of epidemiology at the University of Pittsburg, whose radiation risk estimates are higher than those of the ICRP, Taylor says flatly, 'Radford is not part of the radiation protection community.'

Though they are at a disadvantage financially, the dissidents have the upper hand in spreading their viewpoint. Predictions of disaster, and accusations of folly in high places, are more likely to attract attention than official reassurances of safety. Taylor expresses the frustration that the radiation protection establishment feels about the power of these 'terror-mongers', as he calls them, when he says, 'the tragedy of this is that a half-dozen of such people can attract more national publicity than the entire remaining recognized and objective profession'. Taylor argues that 'Gofman, Tamplin, and Bross [Dr Irwin Bross, another vocal critic of radiation safety standards] must have cost the taxpayers some millions of dollars by being allowed to pursue frivolous charges and claims'. Ralph Lapp, an American physicist and consultant to nuclear utilities, blames the media for amplifying the voices of a few critics. 'I despair of going through the media,' he told me. 'A few scientists, like Dr John Gofman and Dr Karl Morgan – I'm not saying they're without qualifications, but I'm saying they are rogue scientists – aberrant. The media reports them as if they represent 50 per cent of the scientific community, and they don't. They represent themselves.'

Lapp has had the unusual experience of being on both sides of the radiation controversy. In 1949 he published *Must We Hide?*, as a counter-attack to *No Place to Hide*, David Bradley's disturbing diary of the Crossroads tests. Later, in the 1950s and 1960s, he was a scourge of the nuclear establishment. His many books and newspaper articles, including *The Voyage of the Lucky Dragon*, were largely responsible for awakening public concern about ionizing radiation in general and fallout in particular. 'For quite a while, I was public enemy number one at the Atomic Energy Commission,' he told me. 'I started writing with the Alsop brothers and wrote a book with them which never got published. Admiral Strauss [the then-head of the AEC] killed it . . . These were fairly exciting times for me in the sense that if you're public enemy number one, people are going to take a lot of action against you . . . And there were nasty things, like denial of income. With the AEC against me, I couldn't be a consultant. I didn't even push it.' Though Lapp's views eventually merged with those of the majority, he says that he 'never became truly *persona grata* with the AEC. It rankled with them, and furthermore too many times I was right, and that they'll never forgive you'.

One thing that still worries Lapp is the secrecy surrounding much radiation-related research, especially where weapons are concerned. 'I have respect for the Department of Energy's health effects people . . . I believe they've done a reasonable job. On the other hand the secrecy that surrounds them because they are a weapons agency is defeating. And that's what I had to fight whenever I wanted to get any biological data [from the AEC]. The reports were out of the health and medicine division, but all they do is mix a few weapons yields in there and claim it's classified.'

Despite his experiences, which included having Herbert Brownell, Eisenhower's Attorney General, sign a warrant for his arrest in late 1952 for publishing an article in the *Saturday Evening Post* about the first hydrogen bomb, based on non-classified information, Lapp doesn't believe today's dissenters are subject to the same persecutions. 'In those days they [the establishment] could do it; today I'd be in a much prettier position. In those days you were alone.' There are, Lapp points out, relatively few dissenters from the majority view, and, in his opinion, they are ostracized because they are poor scientists, not because they are threats to the established order. 'Karl Morgan is a gentleman, but ... an aberrent scientist'; 'Gofman's estimates are simply weird ... plus there are some things that he doesn't know about fallout'; 'Mancuso is a politician ... very oddball, and highly oriented along union lines.'

The ICRP and its supporters 'see all its critics as tools of the anti-nukes', says Tom Cochran, a staff scientist with the Natural Resources Defense Council. Lauriston Taylor, for example, lumps 'Morgan, Nader, Fonda, etc.', all together as opponents of nuclear power. In reality, the scientists who criticize the official risk-estimates have a range of opinions on nuclear power, nuclear weapons, and the hazards of radiation. John Gofman supports a nuclear defence. Karl Morgan, though very critical of existing safety standards, is not opposed to nuclear power. 'I'm for the safe development of nuclear power,' he told me, 'and I resent that the nuclear industry has often resorted to half-truths and cover-ups ... Reactors could have been built to be much safer.' Despite their misgivings, none of the critics would ban the use of radiation; most have worked with it themselves. 'There are many hazards in life far more serious than radiation,' says Gofman. 'A lot of people are over-worried about radiation as a hazard.'

In an editorial in *Science* magazine, Philip Abelson, a nuclear chemist and advisor to the AEC, acknowledged the difficulties faced by dissenters. 'In questioning the wisdom of the establishment, [the scientist] pays a price and incurs hazards. He is diverted from his professional activities. He stirs the enmity of powerful foes. He fears that reprisals may extend beyond him to his institution. Perhaps he fears shadows, but in a day when almost all research institutions are highly dependent on federal funds, prudence seems to dictate silence.'

The psychological stress of being ostracized is considerable. 'You're up against a pretty unyielding bureaucracy. No matter what you say, they will find a way of making you wrong,' says Victor Gilinsky, who as a member of the US government's Nuclear Regulatory Commission from 1975 to 1984 was himself part of that bureaucracy. 'It's a strong system. People fighting against it tend to be obsessed. Ordinary people go on to something else. It takes a very stubborn individual to last it out.' Some dissenters have become obsessed with justifying themselves. A few have returned to the fold, rather than endure exile. Others seem philosophical about the personal attacks and the loss of livelihood. Perhaps this attitude is most common in older scientists, those who achieved distinction before being labelled as rebels. Joseph Rotblat and Karl Morgan are both 79. Alice Stewart is 81 years old.

Alice Stewart is one of the ICRP's best-known critics. The daughter of two doctors, she entered Cambridge in the 1920s as one of only five women then studying medicine. 'I learned then what it was to be in the minority,' she recalls. She learned it again when she published her famous 1958 paper linking childhood cancers to X-rays. Her conclusions were disputed with diminishing intensity for several decades. Today they are universally accepted. Her more recent work, on the Mancuso–Hanford study and on the relation between naturally occurring radiation and cancer, is still not generally accepted, and may never be. But she is sure that time will prove her right. 'Twenty years,' she says, confidently. 'New ideas take a generation.'

Karl Morgan is often referred to as 'the father of health physics'. During the war he was responsible for the radiation protection of scientists and technicians working at Oak Ridge, the Manhattan Project's plutonium production plant in Tennessee. Afterwards he was head of the health-physics section at Oak Ridge for 29 years. He co-founded and was the first president of the Health Physics Society. He founded the International Radiation Protection Association. He has been a member of the ICRP since 1959. As chairman for 14 years of the ICRP's committee on internal dose, and for nine years of the equivalent committee of the NCRP, Morgan, perhaps more than any other individual, was responsible for setting the postwar limits on internal emitters, the standards that are meant to protect us from the worst effects of radiation.

In 1964, Morgan criticized what he regarded as the medical profession's cavalier use of X-rays, saying that much medical exposure was unacceptable because it was unnecessary. In the mid-1970s, he said that the permissible doses of radiation should be halved. He has often criticized the ICRP's actions. Morgan is now a scientific pariah to most of his former associates. Lauriston Taylor, his longtime colleague on both the ICRP and the NCRP, says that Morgan 'has no reputation to lose'. David Sowby says, 'He takes a different line now. I think he's gone "round the bend".' Taylor has also suggested that Morgan is in it for the money. 'There's great financial gain in being an expert witness for these [radiation injury] cases . . . It's a potential motive,' says Taylor, who has faced Morgan as an expert witness in at least one lawsuit.

Morgan's former employer, the US government, has successfully communicated its low opinion of him to at least some laymen. In Johnston v. the United States, a radiation injury suit against the US government, Federal Judge Patrick Kelly accepted the government's criticisms of Morgan, who appeared as an expert witness for the prosecution. In his 1984 written decision finding against the plaintiff, Kelly said, 'Dr Morgan is perhaps an esteemed scientist of yesterday, trying to hold on to whatever reputation remains . . . He is, in the court's view, a pathetic figure who can better serve the field by simply "going home". Dr Morgan's testimony is stricken from this case as totally unreliable.' Kelly had more faith in Taylor and the experts who appeared on behalf of the US government. He called them 'superb', 'eminent', 'wholly effective, honest and reliable', 'realistic and sound', 'impressive' and 'brilliant'.

Morgan has many defenders, but they too are outside the establishment. 'In my opinion,' says John Gofman, 'he's close to a saint because he's a human being and a top-flight scientist.' Tom Cochran, a former student of Morgan's, says, 'Karl Morgan sees himself as assisting the underdog, those who can't do the calculations and need help. Others could twist that around as him attacking the industry and getting paid for it. But Karl could make so much more playing their game. A great deal of his work is done for free . . . You certainly don't go into environmental litigation if your objective is wealth.'

Morgan himself shrugs off some of the criticism. 'I still consider Laurie Taylor a friend,' he says, adding, 'There are some friends that do things that you hate.' But he also speaks of 'those that are on the ICRP and the NCRP that are obligated to the Department of Energy, and the Nuclear Regulatory Commission, to the nuclear industry. Consciously or sub-consciously, perhaps unintentionally, they are biased in their views'. Says Morgan, who is semi-retired, 'I don't have to please the boss. I don't have to leave out things in my research to get the next raise, or to hold my job, or to be on the list of candidates for a prize.'

Morgan and others criticize the make-up of ICRP. Since its founding by an international group of radiologists, the ICRP has been closely tied to users of medical radiation; about half the commission's members have been medical doctors. Physicists have been the second-largest group on the ICRP. Only a handful of biologists, geneticists, or specialists in environmental or occupational health have ever served on the commission. At present, the commission includes six doctors, six physicists, and one biophysicist. Ten of its members are employed by government agencies, one by the United Nations, and two have university posts. 'A large fraction of members of ICRP have been and are radiologists, employees of the nuclear establishment or have large contracts with the nuclear establishment,' says Morgan, who himself spent more than thirty years working for the AEC and other agencies of that establishment. 'In spite of its usefulness in the past, the ICRP has never been willing to offend the establishment and I'm not sure it's an organization I would trust with my life.'

The public has its own way of deciding which risks are acceptable, and which are not. The results are often in conflict with the views of those who are supposed to know, the experts. There is a tendency, many experts believe, to put undue emphasis on making certain relatively safe occupations even safer, while more dangerous pastimes are tolerated without complaint. In particular, it seems that people are much more concerned about risks connected with radiation than they are about many activities that most researchers say are actually more dangerous, like smoking cigarettes and driving automobiles. The public's concern, according to one health physicist, is 'out of proportion to the best available quantitative estimates of the risk'.

'Our society expects a higher standard of protection against radiation than against almost all other dangers,' says John Dunster. Health concern about radiation, Dunster argues, has been exaggerated 'to anxiety or even phobia'. The British government, for example, has recently spent almost $500 million to reduce radioactive discharges from the Sellafield (formerly Windscale) reprocessing complex – 'probably more than we have spent on all other environmental measures', according to William Waldegrave, the minister responsible. Even after the reductions, Sellafield is still the most-polluting reprocessing plant in the world, but Dr Donald Avery, the former deputy managing director of Sellafield's parent company, argues that the money would have been better spent building 'a number of hospitals which would, I think without question, save more lives'. A November 1985 editorial in the in-house journal of the National Radiological Protection Board concurs: 'The few deaths which are attributed to waste disposal operations will occur in a population that will experience some 10,000 million deaths in the process of renewing itself perhaps a hundred times over the relevant period ... Surely at this level of individual impact we must ask ourselves the question: 'Are we being conscientious or just plain silly?' In the United States, government analysts have an unofficial rule of thumb that says averting one person-rem of exposure should cost no more than $1,000. Using ICRP risk estimates, that means that averting one fatal cancer per year would cost $10 million. By comparison, it is calculated that it costs only $40,000 to save a life with smoke detectors, and only $3,000 to save a life with seat belts.

In 1984 the US Department of Energy awarded an $85,000 contract to Robert DuPont, a psychiatrist and president of the Phobia Society, to find ways to overcome the public's 'nuclear phobia'. DuPont, and others, argue that the public's view of the facts is distorted by 'shock-horror' stories in the press, which in turn is influenced by anti-nuclear ideologues. A well-informed person who approached the issue logically would fear cigarette smoking and automobile driving more than nuclear power plants, they argue. William Allman, writing in the magazine *Science 85*, paraphrased the view of the experts thus: 'In short, we, the general public, are irrational and uninformed. We don't understand probability, are biased by the news media and have a fear of some technologies that borders on the primeval.'

Is the public really so irrational? Some analysts argue that the public's response has a deeper logic than that of the professional risk-experts. It appears, first of all, that people are willing to accept higher risks in activities that they choose, such as cigarette smoking or flying, than in those over which they have no control, such as eating food additives or inhaling emissions from nuclear waste dumps. Another unattractive feature of many such corporate or societally-imposed risks is that they – and their associated benefits – are not evenly distributed among the population. Lauriston Taylor concedes that 'it is only in the use of medical procedures involving radiation that the risk, if any, is compensated by some benefit to the person at risk'. People also seem more inclined to risk sudden death – in a fall or an automobile accident, for example, than a lingering and extremely painful illness like cancer, whether or not it is fatal. As one MIT professor told the *New York Times* shortly after the Chernobyl accident, 'People hate to contemplate dying in strange ways. This makes a thing like radiation seem all the more menacing.'

The public also judges risks upon the basis of its confidence in what the authorities say about those risks, and the probable consequences if the authorities are proven wrong. Public confidence in nuclear official-dom has been declining for the past 30 years, largely due to the public's repeated discoveries that it has been lied to. Lauriston Taylor's argument 'that at the time information was concealed or distorted, there were considered to be acceptable political or economic reasons to warrant it' offers no comfort to those who fear that they still are not getting the whole story. Not surprisingly, the more controversial the subject and the more uncertain the degree of risk, the more the public is likely to require positive proof of safety, rather than being satisfied with the absence of evidence of danger. This is especially true if an accident, however unlikely, could cause damage on a global scale, or injure future generations. It appears that many people can accept a higher risk of a relatively localized chemical accident, even one as terrible as the gas leak in Bhopal, India, that killed at least 2,000 people and seriously injured tens of thousands more, than a Chernobyl-type nuclear accident whose immediate death toll was 31, but whose long-term and global consequences are unknown. 'Our fears,' remarks Allman, 'may tell us a lot about how a risk affects society as a whole.'

CHAPTER EIGHTEEN

1977 Standards

The Bear, Woodstock, Oxfordshire, is a picturesque half-timbered inn, formerly a staging post for coaches carrying mail from the capital to the west of England. It now welcomes busloads of foreign tourists who are being shepherded to nearby Blenheim Palace, ancestral home of the Dukes of Marlborough and birthplace of Winston Churchill. In January 1977, the ICRP held a week-long meeting in this ancient hostelry to finalize the first major revision of radiation protection standards since 1956. The commission faced a difficult task. Adopting new protection guidelines would require changes in radiation equipment and working procedures. Before the war, such changes would have affected relatively few laboratories and hospitals, but since the mid-1950s a worldwide nuclear industry with far-reaching political and economic influence had developed – and developed so rapidly that any significant changes would be awkward and possibly prohibitively expensive.

In this revision the ICRP was concerned more with philosophical than with scientific questions. Unlike earlier revisions, this one was prompted not by new ideas about the risks of radiation exposure, but by new views on the acceptability of those risks. The very first radiation standards – adopted in 1934 – aimed to ensure the safety of radiation users. In 1948, however, radiation protection specialists formally acknowledged what they had long realized – that no amount of radiation was absolutely safe. The new aim of radiation protection became to ensure that the benefits of radiation – national security, cheap energy, efficient medical treatment – would outweigh the risks of the exposures its use entailed. In 1977, the commission took its philosophy one step further by setting a ceiling on risk. It declared that radiation work should not be more dangerous than 'other occupations recognized as having high standards of safety'.

Having studied 'safe' industries, such as manufacturing, the commission had decided that a 1-in-10,000 risk of accidental death per year is acceptable for radiation workers. The accidental death rate for the manufacturing industry in 1986 was 0.33 deaths per 10,000 full-time workers. 'It appears that industries with a death rate of 1-in-10,000 are about as dangerous as the public will accept,' says David Sowby, the commission's former scientific secretary.

Many radiation jobs, of course, also entail some risk of ordinary industrial injury. No one keeps statistics on the accidental death rate among nuclear power plant workers, but workers in the electrical service industry, which includes nuclear power workers, have an annual accidental death rate of about 1-in-10,000, according to the US National Safety Council. If these accidents were considered there would be no room for radiation-related injuries, but the ICRP does not take account of 'ordinary' accidents in calculating the injury rate of radiation workers.

The ICRP's decision that the annual death rate for radiation workers should not exceed 1-in-10,000 created a problem. It meant that the existing dose limit of five rems per year, adopted in 1956, was too high. A five-rem exposure, according to the commission, carries a 6-in-10,000 chance of dying from cancer, plus a 4-in-10,000 chance of passing on serious genetic effects to future generations, not to mention a roughly 1-in-1,000 chance of developing a non-fatal cancer. The commission concluded, says Sowby, that 'it wouldn't really be tolerable to go on exposing somebody at or near the dose limit year in and year out'. For the risks of radiation work to equal those of 'safe' industries, the five-rem limit would have to be divided by ten.

The ICRP's reasoning brought it close to the position taken by many of its critics, who had been arguing that it was both necessary and practical to cut the five-rem limit by a factor of two, five, or ten. Already many countries have succeeded in bringing workers' radiation doses significantly below the five-rem level. The US Environmental Protection Agency's records indicate that 97 per cent of all US radiation workers now receive less than one rem per year. According to the EPA's latest figures, medical radiation workers receive an average dose of 150 millirems per year; flight crews average 170 millirems per year and industrial radiographers average 430 millirems. In fact, the only group in the US whose average annual exposure exceeds 500 millirems are workers at nuclear power plants, who average 650 millirems per year.

Employers, however, argued against lowering the limit on the grounds that doing so would make certain essential jobs impossible. 'If we reduced the limit by a factor of ten, you could not in effect run anything, not a nuclear power plant, not a medical facility, not anything,' says a spokesman for the Atomic Industrial Forum, the nuclear power industry's trade association. (After Chernobyl, the group rechristened itself the US Council on Energy Awareness.) Even a smaller change would make certain key jobs prohibitively expensive or impossible to perform, according to industry and government officials. A 1986 EPA paper reported the nuclear industry's conviction that 'a five-rem annual limit is needed for a small number of highly skilled workers doing such specialized work as steam generator repairs' and that 'there is a further need for exposures above five rem/yr for a small number of skilled workers involved in nuclear power plant service'.

Among those whose jobs are known to expose them to above-average levels of radiation are maintenance workers in power plants, industrial radiographers (people who use X-rays to check the integrity of welds, for example), and some medical personnel. Some of these high-exposure jobs require no special training. In those cases, the job – and the dose – can be split among a number of unskilled, temporary workers. However, for many of the more specialized tasks, 'hiring or training qualified replacement workers is difficult or impossible,' according to the Atomic Industrial Forum. The American College of Radiology warns that reducing the five-rem limit could restrict the frequency with which some professionals, such as cardiologists, are able to carry out certain critical medical procedures. The EPA estimates that lowering the limit to one or two rems would cost hospitals several hundred million dollars a year in improved monitoring, training, and redesigned radiation equipment.

Former NRC commissioner Victor Gilinsky believes that the nuclear power industry could comply with exposure standards as low as one rem per year. 'They'd have to work a lot harder at it, but exposures in other countries are quite a bit lower than they are here, not by a factor of ten, but by quite a bit . . . Where they'd get into trouble is not so much operating plants, but repairing them. You'd have to have a lot more workers.' The problem is that many nuclear reactors have deteriorated more quickly than experts predicted. One result is that a great deal of hazardous maintenance and repair work is necessary to keep deteriorating plants functioning. A survey by the Electric Power Research Institute found that, worldwide, two-thirds of pressurized water-reactors, the most common type of reactor, have corroded tubes in their steam generators, leaking radioactive water. These problems mean that reactors require constant maintenance and repair, jobs that entail a significant degree of radiation exposure.

The Atomic Industrial Forum estimates that a reduction in the exposure limit from 5 to 0.5 rems per year would cost the average nuclear power plant more than $6 million extra annually. According to the Nuclear Regulatory Commission, such a move would eliminate 2.6 million person-rems of exposure between 1980 and the year 2000, thus eliminating, according to ICRP risk estimates, 325 cancer deaths and more than 200 genetic defects. A one-rem limit would cost each power plant an additional $1.5 million each year, according to the forum. A survey by the Environmental Protection Agency found that an annual limit of 1.5 rem, would require the hiring of almost 30,000 new radiation workers by the nuclear industry as a whole, at a cost of between $400 million to $700 million. A 1.5-rem limit, an AIF spokesman said, 'would severely restrict specialty workers to the point of making certain tasks difficult or even impossible to perform'.

It has been suggested that the limit be lowered to one rem or less for the vast majority of workers with low-exposure jobs, and that exceptions be made for certain types of essential, high-exposure work perhaps restricted to workers who have passed child-bearing age. The EPA has considered and rejected that idea – as has the nuclear industry, which fears the psychological and legal impact of a double standard. 'I think it would be very bad to pick out a class of workers and say, you've got special skills and we want them, and we'll pay you more because you're over 45 and you're no longer going to have kids,' remarked the forum's Dave Harward. 'Let's suppose that you offer somebody more money or more vacation or something to take that risk. That person could skin you because that's inequitable treatment. You presumably know that one of those jobs might cause him problems and you let him do it anyway.'

In its 1977 recommendations, known as ICRP–26, the ICRP tried to chart a middle course between the nuclear industry's fears and its own conclusions that the existing five-rem dose limit was ten times too high. In Sowby's words, 'The commission didn't adopt a new number, it adopted a new concept.' The concept is that, although five rems is still the legal dose limit, it is no longer an acceptable level of exposure. Instead, says the commission, 'all exposures shall be kept as low as reasonably achievable, economic and social factors being taken into account'. This is known as the ALARA concept (ALARA being an acronym of 'as low as reasonably achievable').

The commission has made it clear that it expects the ALARA principle to bring the average worker's exposure down to no more than 0.5 rem (500 millirems) per year. As Sowby explained, 'our experience has shown that when you set a limit, the actual dose will tend to be about a tenth of that. And it's that tenth that is the acceptable limit.' The legal limit, however, is still five rems per year.

'ALARA says, "If it doesn't cost an unreasonable amount to do it, then you should avoid radiation exposure,"' explains Michael Thorne, Sowby's successor as the ICRP's scientific secretary. The 'reasonableness' of reducing exposure will vary from industry to industry and from country to country according to the prevailing economic and social conditions. 'It's a matter of national and local judgement as to how much it's worth spending to reduce risk,' points out Thorne. 'Take the Third World, for example. It's not necessarily appropriate to spend the same amount of money on reducing radiation risks in an environment where there may be many other competing risks that are clamouring for a small amount of government resources.'

The decision as to whether or not a lowered exposure is 'reasonably achievable' should be based, says the commission, on a cost-benefit analysis. This procedure weighs the costs of introducing safeguards against the benefits that would result from their introduction. Though the idea of cost-benefit analyses is to make decision-making more objective, the technique itself requires many subjective judgements. In order to compare like with like, analysts must translate all costs and benefits into financial terms. This means attaching monetary values to factors as simple as buying new monitoring equipment, or as complex as saving a human life. Though many people find assigning a value to a human life abhorrent, cost-benefit analyses require it. Various US government agencies have set the value of a human life at levels ranging from $650,083 to $7 million. Many papers have been written on the subject of how to price a life, or the loss of one, whether to include a subject's projected future earnings, the cost of his or her medical treatment before death, the industrial cost of days lost on the job, the cost to the state of supporting a fatherless or motherless family, and so on. Deciding what costs and benefits to include in the calculations can also be difficult. The ICRP disregards non-fatal cancers (which are roughly one-and-one-half to two times as common as fatal cancers), but many critics say that they should be included as 'costs' of radiation exposure. Since certain types of cancer, such as skin and thyroid cancers, are rarely fatal, concentrating exclusively on cancer deaths means that these painful diseases are almost completely disregarded.

The ALARA principle has been criticized as ambiguous by people on all sides of the nuclear debate. Critics of the radiation establishment say that it is so vague as to be unenforceable. Cost-benefit analyses are not yet so precise that they can yield a definitive answer about whether or not something is 'reasonably achievable'. In reality few decision-makers actually understand or use them. The final decision on whether or not to introduce more stringent controls is more likely to be influenced by political pressure than by mathematical equations according to Sir Kelvin Spencer, former chief scientist for Britain's Ministry of Power. 'We must remember that government scientists are in chains,' he said. 'Speaking as a one-time government scientist, I well know that "reasonably achievable" has to be interpreted, so long as one is in government service, as being whatever level of contamination is compatible with the economic well-being of the industry responsible for the pollution under scrutiny.'

On the other hand, industry representatives fear that ALARA will trigger an expensive and pointless downward spiral of exposure limits, necessitating increasingly burdensome administrative arrangements. Lauriston Taylor, for one, is concerned that, even if exposures are very low, the ALARA principle will encourage 'somebody to come along with another yank on the ratchet. There's no end to this. I don't know anything about the economics of this whole question of radiation, but it's obvious that you can strangle yourself. You can do yourself out of the usefulness of radiation, in spite of the fact that radiation is probably the safest of a vast number of carcinogenic-related industries.'

In fact, concern about the ALARA principle was one reason why the ICRP and its American sister-body, the National Council on Radiation Protection, were out of step for an unprecedented ten years. The NCRP is an independent advisory body that performs essentially the same job as does the ICRP, although its recommendations are specifically for American users of radiation. The two groups have an overlapping membership and have always been very close, making essentially the same recommendations within a year or two of one another. But it took the NCRP ten years to come into line with the international commission's 1977 recommendations. NCRP's version, Report 91, was delayed by internal controversy over, among other things, ALARA.

'There's an awful lot of people who feel that there's just no justification for any change in radiation protection exposure limits,' says William Beckner, the NCRP staff scientist who helped prepare Report 91. 'They say, "It's down low enough; you don't have any reason to believe that there is any significant number of people who are suffering any detrimental effects under the present regulations, so why do you want to change?"'

Wishing to stay in step with ICRP, however, NCRP finally did adopt ALARA, but tried to put a brake on it by adopting yet another new principle, the principle of 'negligible individual risk level' (NIRL), also known as *de minimus* dose, from the Latin phrase, *de minimus non curat lex*, meaning 'the law does not concern itself with trifles'. According to this concept, if a person's exposure to any single source of radiation is – or is anticipated to be – less than one millirem (one-thousandth of a rem) per year, it need not be monitored or otherwise considered. 'You can just forget it completely,' Beckner explains. Even assuming a person was exposed to ten such negligible sources – a highly unlikely situation, says Beckner – the maximum disregarded exposure would be ten millirems. The point of NIRL, says Beckner, is to provide 'a bottom-line for ALARA; otherwise you'd just keep pushing it further and further'.

Besides the ALARA concept, the main change introduced by ICRP–26 has to do with internal exposures. ICRP–26 established a system of 'weighting' the main internal organs as a proportion of the whole body, so that internal and external exposures could be added together. This was a logical improvement over the old system, under which internal and external exposures were considered separately, as if they could not happen to the same person. The new system is controversial, however, because the formula used permits higher doses to the major internal organs – the first time in radiation protection history that limits have been relaxed rather than tightened. The annual limit for the gonads was raised from 5 to 20 rems; for breasts, from 15 to 32 rems; for the bone marrow, from 5 to 42 rems; for the lung, from 15 to 42 rems; and for the thyroid and bone surfaces, from 30 to 167 rems. Because 167 rems was regarded as likely to cause symptoms of radiation sickness, the commission arbitrarily lowered the exposure limit to the thyroid and bone to 50 rems.

Using new computer models of how radioactive materials behave inside the body, the commission also revised the intake limits of 260 radionuclides – the amounts of them that workers may ingest or inhale without exceeding internal exposure limits. Some intake limits were lowered but the intake limits for about 170 radionuclides were raised. The permissible level of iodine–131, for example, is increased by eight times, and that of strontium–90 by 17 times.

The commission has defended itself against charges of having relaxed protection standards by saying that, although raising the limits made biological and mathematical sense, the ALARA principle should prevent employers from allowing workers to actually receive those higher doses. Nonetheless, warns Robert Alexander of the US Nuclear Regulatory Commission, 'in many cases the concentration of radioactive material can be raised and undoubtedly will be raised by our licensees'. Robert Alvarez, of the Washington DC-based Environmental Policy Institute, argues that workers in Third World countries will suffer most under the new limits: 'They say, "We don't intend the worker to actually get these doses," that employers should keep exposures below the legal limits . . . But what about India, what about the Philippines?'

Regarding the new internal standards, Alexander says, 'I was a critic and I still am. Where you have factories and laboratories where people have worked successfully with a given radionuclide for 20 years and have proven that they can make an adequate profit and run their business successfully with the present limits, then I can find no justification for changing the laws in such a way that more radioactive material can be introduced into the worker's body.'

In the United States, pressure both from the public and from the huge civil and military nuclear industry has caused the government to make its own rules – sometimes more stringent, sometimes the reverse – on certain points. For example, rather than accept the ICRP's recommendation that pregnant radiation workers can receive up to 1,500 millirems of radiation, the NRC has proposed restricting pregnant workers to 500 millirems. The NRC also wants to adopt the NCRP's idea of a *de minimus* dose as a lower limit to the ALARA principle.

Overall, however, the NRC's proposed rules stick fairly closely to IRCP–26. Former NRC commissioner Victor Gilinsky recalls that the commission was 'anxious to be in tune with the international body . . . We're dealing in an area where there aren't many hard facts, so you have to rely on judgement to a large extent. And particularly in governmental decision-making, it's nice to be able to say that you are relying on some unimpeachable source . . . somehow it gives the whole thing more respectability.'

ICRP-26 also concluded that a 1-in-100,000 to 1-in-1,000,000 annual risk of premature death due to radiation 'would be likely to be acceptable to any individual member of the public'. This implies that a member of the public should not average more than 100 millirems per year. The commission declined, however, to lower the existing limit of 500 millirems per year. Instead, it expressed confidence that average exposures would not go above 100 millirems per year. The commission went on to say that if the average exposure of members of the public does exceed 100 millirems per year, 'the situation might still be justifiable, even though the average risk for members of the public would be higher'.

The ICRP's decision to retain the 500-millirem public limit, but to urge that exposures be kept well below that level, was intended to accommodate those industries that said they could not comply with a stricter limit – at least not without going bankrupt. US regulators have followed the same path, setting different standards for different types of nuclear facilities according to their ability to comply. Each 'uranium fuel cycle' operation, for example a uranium mill, fuel fabrication plant, or nuclear power station, can expose people living near by to 25 millirems of external radiation a year. If two or more uranium fuel cycle facilities are within ten miles of one another, the dose from their combined releases must not exceed 25 millirems. By contrast, certain other radiation facilities, including industrial plants that use radiation, hospitals, research labs, and high-level radioactive waste dumps, can legally expose people to 500 millirems of external radiation per year. If there are several such facilities in one area, they can each deliver the maximum dose. The US Nuclear Regulatory Commission is proposing to insist that these facilities not expose their neighbours to more than 100 millirems a year, but any that cannot meet the new limit would be allowed to continue giving doses of up to 500 millirems a year.

Regulating these facilities more strictly, more in accordance with the spirit of ICRP-26, would be too expensive, say officials. 'One of the main reasons why we can't reduce the limit further, why we have to leave an exemption in there, are the radiation therapy units,' says Hal Peterson, a senior health physicist in the NRC's Division of

Regulatory Applications. Most medical and industrial radiation units use cobalt–60, which emits highly penetrating gamma rays, as their radiation source. Hospitals, particularly those in older buildings, find it especially difficult to isolate or shield their cobalt sources sufficiently to guarantee lower exposures. The NRC does not try to control multiple exposures from by-product facilities because, in Peterson's words, 'there are so many overlapping sources that it would be a nightmare to regulate'. Nonetheless, Peterson asserts that the double standard for public emissions does not mean that the public is receiving too much radiation from some nuclear facilities. The 25-millirem limit for certain nuclear plants 'derives from a cost-effectiveness analysis, not from a health analysis,' he says.

Nuclear weapons facilities, which are owned and regulated by the Department of Energy, are also allowed to expose members of the public to 500 millirems per year – 100 millirems per year if the exposure is to last for more than five years. The DoE owns 18 major nuclear facilities and many smaller ones scattered around the country. The plants are operated by private contractors, including Dupont, American Telephone & Telegraph, Monsanto, the University of California, and the Massachusetts Institute of Technology. Many of these facilities were designed and built when radiation standards were many times less stringent than they are today. 'The design basis for the DoE's operations is literally in the 1940s and early 1950s,' says Robert Alvarez of the Environmental Policy Institute in Washington DC. 'The nuclear weapons bureaucracy is most sensitive about reducing exposure, because they would have to rebuild their entire industry.'

The Hanford nuclear complex's plutonium-producing N reactor, inaugurated in 1963, was designed to have a 20-year operating life. In common with the Chernobyl reactor, the N reactor lacks the thick containment walls that surround modern nuclear power plants. In 1986, the 23-year-old plant was still operating and, despite protests that it was dangerous, the DoE planned to keep it going for another 15 years. According to Energy Under-secretary Joseph Salgado, the United States weapons' programme needed the plutonium that the reactor could produce. 'National security reasons do not allow the permanent shutdown of the N reactor,' he told journalists. The N reactor was closed after the Chernobyl disaster for repairs so that it could continue operating until 1991, when, the DoE said, unsafe conditions at the plant would be irremediable. In February 1988, however, congressional pressure led the DoE to permanently close the

N reactor. The DoE has five other plutonium production plants – all at the Savannah River Plant in South Carolina. Two of them are closed down, and the other three were reduced to half-power in early 1987 after the National Academy of Sciences warned that their emergency cooling systems are inadequate. In 1986, the DoE also temporarily shut Hanford's Plutonium Uranium Extraction plant, nicknamed PUREX, and a sister plant, shortly after an audit revealed that plutonium was being stored in an unsafe manner, and found some equipment in sub-standard condition. In 1987, five large reactors at Oak Ridge National Laboratories were closed for safety reasons.

The ICRP's public exposure limit is supposed to include all non-background radiations except those from medical treatments. In 1977 the commission issued a warning about the dangers of multiple exposures, pointing out that 'an increase in the average dose to members of the public could result from any large increase in the number of sources of exposure, even though each . . . caused no exposures above the recommended limits'. In practice, however, most governments treat each radiation facility as though it were operating in isolation, allowing each one to emit up to the limit.

In 1985, the commission took a slightly firmer line on the public exposure limit, saying 'the commission's present view is that the principal limit is 100 millirems in a year', but that industries that cannot meet this limit may still use 500 millirems 'for some years'.

PART FOUR

CHAPTER NINETEEN

Natural Radiation

Radiation exposure is nothing new. Every person who has ever lived has been constantly bombarded by ionizing radiation from conception to death – and long after. Each of us is exposed to radiation from outer space, from the earth, and from our own bodies. There is a mistaken tendency to assume that natural radiation is harmless. Ionizing radiation is a hazardous phenomenon whether it is natural or man-made. A certain – unknown – proportion of the cancers and genetic defects the human race suffers are the result of exposure to naturally occurring radiation. British researchers Alice Stewart and George Kneale have recently found correlations between varying levels of natural background radiation in the United Kingdom and the incidence of leukaemia in children. They believe that natural radiation may be the major cause of childhood leukaemias. Most scientists believe that natural radiation causes about one per cent of all fatal cancers.

In just the time it takes you to read this sentence, several hundred cosmic rays will have shot through your body. Cosmic rays are highly penetrating ionizing radiations that originate mostly in outer space, though during periods of solar flares, some cosmic rays also come from the sun. Cosmic rays irradiate the earth and everything in it. As they move through the atmosphere, they also interact with elements in the atmosphere to produce secondary radiations and new radionuclides. Most of these so-called 'cosmogenic' radionuclides are short-lived, but carbon–14 and hydrogen–3 (commonly called tritium), survive long enough to have become incorporated into the environment and our body tissues.

The earth itself contains radioactive materials that were formed at the birth of our universe, billions of years ago. At that time the earth was even more radioactive than it is today. Most of the primordial radionuclides have decayed out of existence by now; only the ones with half-lives of 100 million years, or more, have survived to the present. About 20 of the surviving primordials decay to stability in a single step.

Of this group, only potassium–40 and rubidium–87 are abundant enough and radioactive enough to have a significant impact on living things. Three primordials – uranium–238, thorium–232, and uranium–235 – lose their radioactivity through a series of changes rather than in a single step. Together, these three 'decay series' comprise about forty radionuclides, all of which make their way into water, food, air, and the human body.

All living things contain potassium, and all potassium contains a small amount of the radioactive isotope, potassium–40. Thus, there is radioactive potassium in all plants and animals, including ourselves. A 150-pound person contains about 140 grams of radioactive potassium, mostly in the muscles. Both because it is widespread and because it is quite radioactive, potassium–40 is normally the most significant radionuclide in food and human tissue.

All rocks and soils contain some uranium and thorium, both of which give off alpha and gamma rays. Igneous rocks have only about 0.03 parts per million (ppm) of uranium; granite has 4 ppm; bituminous shale has up to 80 ppm. The highest concentrations of uranium are found in phosphate rock, some of which contain 120 ppm of uranium. The decay products of uranium and thorium, including radon gas, are also present to some degree in all rocks and soils. A certain amount of radon escapes from rocks and soil to the atmosphere; eventually its radioactive daughter products fall to the ground, contaminating the vegetation on which they are deposited. People and animals who consume large quantities of leafy plants are exposed to quite high levels of these radionuclides. Tobacco smokers and the thousands of Laplanders and Eskimos whose diet consists largely of reindeer that feed on lichens are especially vulnerable. Both these groups have high levels of radioactive lead–210 and polonium–210 in their bodies.

Some radionuclides, depending on how readily they dissolve, are carried by rain into drinking water supplies. Radium dissolves easily in water. In 1982, a study found that drinking water supplies in several sizable towns in Illinois, Iowa, Missouri, Wisconsin and the Middle Atlantic states, contain radium at concentrations that exceed the Environmental Protection Agency's proposed limit on radium in drinking water of five picocuries per litre. Showering and washing with radium-rich water releases radon gas into the home. In his textbook, *Radioactivity in the Environment*, Ronald Kathren, a health physicist at the DoE's Battelle Pacific Northwest Laboratories, warns that 'the potential problem of radon released from domestic waters has as yet received scant serious consideration'.

The average person receives a dose of about 100 millirems a year from this background radiation. About one-third of this is external exposure from cosmic radiation. One-third is external exposure from gamma radiation emitted by radioactive elements in the earth's crust. The final one-third comes from exposure to radionuclides that exist naturally in the body, like potassium−40, or those that are inhaled or eaten, like uranium and thorium and their decay products. One hundred millirems per year is only an average dose; individual doses vary widely, according to local conditions and each person's habits.

Naturally occurring radiation is not evenly distributed. Cosmic rays, for example, are gradually absorbed as they pass through the atmosphere. Thus, people living at sea level get a smaller dose than do those living at high altitudes. People living in Denver, 5,760 feet above sea level, are exposed to nearly twice as much cosmic radiation as people who live at sea level. The dose at the top of Mount Everest is about 2,000 millirems per year, compared to the sea level dose of approximately 30 millirems.

Levels of terrestrial radioactivity also vary from place to place, largely according to the amount of uranium and thorium present in the local soil and water. Many mineral springs all over the world have quite high concentrations of radium and radon. When this was first realized, some people attributed the traditional curative powers of mineral baths to their newly discovered radioactivity. Underground galleries were carved out, so that patrons could sit in tunnels and breathe in the radon-laden air. Even today, some resorts boast of the high radiation levels of their drinking and bathing waters. A substantial industry has developed along the Atlantic coast of Brazil to cater to tens of thousands of vacationers who come seeking the supposed health benefits of the area's radioactive black sand beaches. The beaches, composed of monazite sand, a mineral naturally high in radioactive elements, are hundreds of times more radioactive than most soils − emitting 17,500 millirems per year at some spots. Radiation levels in the streets of Guarapari, a major black sands tourist town, range from 800 to 1,500 millirems per year. Areas of high natural radioactivity have also been found in the Indian state of Kerala, China's Guangdong Province, the Soviet Union, France, and Madagascar. Another famous radioactive site is Morro do Ferro in the Brazilian state of Minas Gerais, where there is a rich deposit of thorium and other radioactive ores. Plants growing there have absorbed so much radiation that they can produce X-ray photographs of themselves.

For millions of years, the level of naturally occurring radiation to which humans were exposed remained unchanged. But gradually over the last few centuries, and especially in recent years, the human race has developed technologies that have increased its exposure to natural radiation. Underground mining, for example, exposes workers to higher levels of the radionuclides that occur in all rocks.

Airplane flights also increase people's exposure to cosmic rays. Someone who flies on an ordinary aeroplane across the Atlantic receives an average dose from cosmic rays of about three millirems. A person making the same trip on Concorde receives only two millirems, because the flight is so much shorter. Occasional solar flares produce cosmic radiation so intense at high altitudes that high-flying aircraft are forced to take evasive action. Astronauts, especially those who remain in space for more than a short time, can receive substantial doses of radiation. During missions that lasted from three weeks to three months, some Skylab astronauts received more radiation than workers are allowed to get in a whole year. Cosmic radiation may be one of the most important limiting factors for future space exploration.

Phosphate rock contains the element phosphorus, an essential nutrient for all living things. Because of this, phosphate is widely used as a fertilizer and as a supplement in livestock feed. But phosphate also has high concentrations of uranium and thorium. As a result, there is now uranium and radium in virtually all categories of food, and in human tissues as well. Scientists estimate that continued use of phosphate fertilizers could eventually double the amount of radium and uranium naturally present in soil. For most people, however, the main source of exposure to phosphate is phosphogypsum, which is produced from recycled phosphate wastes and used in building materials as a substitute for natural gypsum. The amount of phospho-gypsum produced in 1977 alone will emit 30 million person-rems of radiation over the millions of years that it remains radioactive.

Potassium fertilizers, also in widespread use, add approximately 3,000 curies of radioactive potassium to United States farmlands each year. Fossil fuels also contain potassium (as well as other radioactive

materials), so burning coal or gas releases a certain amount of radioactivity to the atmosphere. When coal is burned, most of the radioactive
material remains in the ash, but some is released to the environment.
Producing one gigawatt of electricity for one year by burning coal in a
power station produces a collective dose commitment of 200 person-
rems. (By comparison, the collective dose commitment from generating one gigawatt-year of electricity with nuclear power is 67,500
person-rems. If the dose from nuclear waste is included, the figure
goes up to 400,000 person-rems). Domestic fireplaces and coal-
burning stoves contribute much more radioactivity to the environment
than do large power stations, even though they consume less coal,
because they lack the power stations' sophisticated pollution-control
equipment.

The rise of radon levels indoors is an important example of how the
human race is increasing its exposure to naturally occurring radiation.
Radon seeps from soil into buildings through their foundations, but
radon-rich water supplies can also contribute to high radon levels in a
home, as can certain uranium-rich building materials, such as granite
and phosphogypsum board. A century ago – before the use of gypsum
board made from uranium-rich phosphate wastes, before builders
knew how to construct weathertight homes, before indoor showers
became popular – radon levels in homes would have been much lower.

Lately, scientists in different parts of the world have been surprised
to find that concentrations of radon in many homes are much higher
than anyone had realized. Surveys in Sweden, Finland, Britain, and the
United States, have found radon concentrations in many homes
hundreds of times greater than those outdoors. Nowadays, in fact, a
person may well receive a radiation dose from radon that equals or
exceeds his or her dose from all other natural sources. In Britain, for
example, the average radon dose is 80 millirems per year, and in certain
areas doses are much higher.

In the United States attention was focused on the problem of indoor
radon in December 1984, when Stanley Watras, an engineer at the
Limerick nuclear power station near Reading, Pennsylvania, set radiation alarms ringing on his way *into* work one day. Officials subsequently
discovered that Watras had been contaminated not by his job, but by
extremely high radon levels in his home, which happened to have been
built on a 30-foot-wide vein of uranium ore. Scientists were amazed to
find that the Watras's home contained 2,700 picocuries of radon per
litre of air – still the highest ever recorded in a building. Watras's

annual dose from radon was estimated to be 40 rems a year – eight times the amount of radiation workers are allowed to receive.

With a half-life of only 3.8 days, radon quickly decays to the longer-lived, alpha-emitting particles, called radon daughters. Radon daughters do not emit penetrating radiation, so they cannot irradiate the internal organs or the sex genes from outside the body. But it is easy to inhale the microscopic particles, and, if they lodge in the lung, they may cause lung cancer. The EPA has proposed a standard for indoor radon of four picocuries per litre of air, which, for a house-bound person, is about one-half the amount of radon to which uranium miners can be legally exposed in a year.

The Environment Protection Agency estimates that radon exposure is the cause of some 5,000 to 20,000 of the 136,000 fatal lung cancers Americans suffer every year. A person who spends 18 hours a day in a house with four picocuries of radon per litre of air over a 70-year lifetime would increase his or her risk of dying from lung cancer by between one and five per cent, according to the EPA. Non-smokers would at least double their normal one-in-one-hundred risk of dying of lung cancer.

Until recently, the average American house was believed to contain roughly one picocurie of radon per litre of air. Though there has been no comprehensive nationwide survey of indoor radon-levels in the United States, high concentrations of indoor radon have been found in homes in 30 states. Some families have been advised to evacuate their homes, and at least one school has been closed due to high radon-levels. The EPA estimates that as many as 12 per cent of the country's 70 million homes may have radon concentrations greater than four picocuries per litre of air. Anthony Nero of Lawrence Berkeley Laboratory estimated in a November 1986 article in *Science* that radon doses in one million American homes 'approximate or exceed those received by underground uranium miners'.

Some of the highest radon-levels in the United States have been found in the uranium-rich Reading Prong, a geological formation that stretches across parts of Pennsylvania, New Jersey, and New York, but high readings have also been found in areas with relatively low concentrations of uranium. Other factors, such as soil permeability, may in some cases be more important than geological conditions in determining how much radon enters a house. Extremely permeable soils, such as coarse gravels, allow radon to travel to the surface, from where it can enter houses. House construction too is a factor; neighbouring homes may have very different readings due to differences in building materials and ventilation. Some realtors predict that potential home-buyers will soon require a radon test before they buy.

The EPA advises home-owners to test for radon using the simple air sampling kits now available in hardware shops or by mail order. The devices, which cost from $10 to $50, must be mailed to a testing firm for analysis. Only New Jersey licenses radon testing companies, but most state public health or environmental agencies have lists of reliable companies. If the radon concentration is higher than four picocuries per litre of air, the EPA recommends a more sophisticated test, possibly followed by remedial measures. Improving ventilation can help reduce the build-up of radon in a building, but that alone will not solve the problem. In fact, an open window, or other form of ventilation, can actually draw more radon into a house by increasing the pressure gradient between the house and the soil. The most important part of reducing indoor radon concentrations is to reduce the amount of radon entering buildings in the first place. The EPA is testing a system for sucking radon out of a building's foundations before it can enter the basement. The Swedes have developed a device called a radon reservoir that they claim can reduce indoor radon contamination over several acres for less than $2,000.

CHAPTER TWENTY

Multiple Exposures

Once solely a medical and research tool, radiation today has hundreds of practical applications. Plant breeders irradiate seeds to create mutant strains of crops. Researchers use radioisotopes to track the movements of fertilizers and pesticides through plants and soils, and to study the global movement patterns of wind and water. Manufacturers use radioisotope gauges to check that beer cans are properly filled and cigarettes are tightly rolled. Doctors use radiation in diagnosis and therapy and to sterilize medical products. Investigators detect forgeries and date natural and man-made objects by measuring radioactive decay. Inventors use radiation to induce chemical reactions that give rise to new products, such as teflon-coated frying pans and super-absorbent disposable diapers. Radioactive materials also go into the making of smoke detectors, self-illuminating watches and instrument panel dials, lightning rods, gas camping lanterns, cellophane dispensers, false teeth, and other consumer products.

No person on earth can escape the multiple exposures of the industrial radiation age. We are all now exposed to radiation from many sources besides nature: fallout, nuclear accidents, medical radiation, radioactive consumer products, nuclear waste, and the production of nuclear weapons and nuclear power. In the United States alone there are today more than 100 uranium mine and mill sites, 72 major commercial power reactors, 280 weapons production facilities (at 20 sites), 3 major high-level radioactive waste disposal sites, 30 nuclear reactors belonging to the Department of Defense and the Department of Energy, 148 naval reactors, 51 weapons storage sites, tens of thousands of places where radiation is used for medical or industrial purposes, and an unknown number of contaminated abandoned sites.

It is becoming harder and harder to isolate nuclear facilities. In Britain, where the government has a policy of trying to site nuclear reactors away from population centres, a remote location is defined as one with fewer than 600,000 people living within ten miles. Moreover, nuclear facilities are frequently clustered together. In the heavily populated San Francisco Bay area, for example, there are more than 500 licensed nuclear facilities, from reactors to large industrial, research, and medical institutions, to laundries that handle the contaminated clothes of nuclear workers. In addition to these civilian nuclear facilities, there are more than ten military reactors and three weapons storage sites. In parts of the southwestern United States, people are subject to overlapping exposures from uranium mines and mills, weapons production, testing, and storage facilities, and nuclear waste dumps.

The radiation protection system, however, is not designed to take account of multiple exposures. 'If you take any one source of exposure, it may not exceed accepted standards, but we found that there are 20 to 25 different sources of exposure for somebody who lives here all their life,' says Helene Hanson, an Indian Health Service nurse working in Arizona's uranium-rich Red Rock Valley. In general, radiation regulations treat each source of radiation as though it were the only one, rather than considering their cumulative impact on individuals or on the population as a whole.

The age of industrial radiation really began with the inauguration – on 2 December 1942, in an underground squash court at the University of Chicago – of the world's first nuclear reactor. Since then, some 1,200 nuclear reactors have been built. Most of them have produced electricity for civil use; the rest have made plutonium for weapons; created radioisotopes and ionizing radiations for industry, medicine, and research; or powered nuclear submarines. All have steadily leaked radioactivity into our air, water, and soil. According to the DoE's Brookhaven National Laboratory, between 1970 and 1981, 50 US commercial reactors emitted 40 million curies of radiation – slightly more than was released in the Chernobyl disaster. In 1980, the public's dose from nuclear power operations world-wide, past and present, was 50,000 person-rems. Assuming that the present level of radioactive discharge is maintained, the collective dose will rise to one million person-rems in the year 2000, and 20 million person-rems in the year 2100.

Most of the radioactive materials released from nuclear reactors and their associated activities – the mining, transport, reprocessing, and disposal of nuclear fuel – are so short-lived that they disappear before they can travel far. These short-lived emissions expose mostly people who live within 50 miles of a nuclear facility. Since they decay so rapidly, they make up almost all the dose that reaches the public in the first 100 years or so. Most of the long-term dose comes from a few long-lived radionuclides – specifically tritium, krypton–85, carbon–14, and iodine–129 – that accumulate in the environment and spread around the globe. People all over the world are exposed to these four radionuclides, but because they decay slowly, most of their dose will be delivered between a thousand and a hundred million years in the future.

According to the United Nations, the nuclear fuel cycle produces about 400,000 person-rems of radiation for every gigawatt-year of electricity produced. That dose will be delivered gradually over the millions of years it will take for all the radionuclides to decay to stability. In 1980, 80 gigawatts of nuclear power was generated, so the dose commitment for that year was 32 million person-rems. The ICRP calculates that a large population exposed to one million rems will suffer 125 premature cancer deaths, and 80 genetic defects. Thus, ICRP statistics indicate that 4,000 people will eventually die, and more than 2,500 will be born with genetic defects, as a result of the 80 gigawatts of nuclear power generated in 1980, though the deaths and injuries will be spread over millions of years. Each additional year of generation at the 1980 level will cause another 4,000 deaths and 2,500 genetic defects. Many scientists believe that these ICRP-based extrapolations are conservative.

Most people who worry about radiation exposure are concerned about the dose they may be getting from large operations, such as nuclear power stations, waste dumps, and weapons facilities. But smaller, seemingly innocuous operations such as hospitals, research laboratories, or industrial plants that generally escape popular scrutiny, may contribute disproportionately to the public dose. People have a tendency to disregard familiar hazards. This tendency may be even stronger in hospitals, labs, and many industries, where the amount of radioactivity handled by any one person is relatively small and where radiation is not the main focus of work, but an adjunct. Policing such operations is awkward and expensive, because there are so many of them – 23,000 in the United States alone, compared to 109 active civil nuclear power reactors. Statistics on exposures from such facilities is mostly anecdotal, since their managers are not required to report their routine releases to any public authority.

Occasionally, however, some hard facts do come to light. In 1987, the California Department of Health Services issued 33 citations for careless handling of radioactive materials to the University of California Medical Center, in San Francisco. The state investigators found that only half of the 950 staff members who handle radioactive material had received formal safety training, and that not all radiation workers wore the requisite film badges to record their exposure. Between 1983 and 1985, the number of sites in the medical complex where radioactive material is used increased by 50 per cent, to 616. Three hundred of these rooms did not carry the required caution signs because lab officials feared they would cause undue alarm to people working or passing near by, or deter firemen from entering during an emergency. The UCSF Medical Centre receives more than 4,000 shipments of radioactive isotopes each year, and generates more than 5,500 cubic feet of radiation-contaminated waste each year. Investigators found that some of that waste had been tossed into a dumpster, poured down sinks, and piled in a public driveway while awaiting shipment to nuclear waste disposal sites.

That these exposures can add up was demonstrated recently in the wealthy commuter town of Weybridge, in London's 'stockbroker belt'. Residents of Weybridge were surprised to learn that they have twice as much radioactive iodine in their body as do people living next to the notorious Windscale reprocessing plant. That fact only came to light when scientists from the Ministry of Agriculture and St Bartholomew's Hospital began looking for uncontaminated populations to compare to those in Windscale. High levels of iodine–125 have also been found in Weybridge's drinking water, which comes from the River Thames, in swans that live on the river, and in cows that graze near by. The iodine is believed to come from research establishments and hospitals flushing their radioactive waste into the sewers.

Just as our descendants for tens of thousands of years will be exposed to the radiations from our activities, so our generation is burdened with the radioactive waste of the past. Many Western states – and one in the East – are plagued with huge piles of radioactive uranium tailings. The tailing piles, some of them the size of large hills, covering hundreds of acres of land, are residues of the milling process by which uranium is extracted from the raw uranium ore. The ore that remains is virtually free of uranium, but is still contaminated with radium, radon, radon daughters, and all the rest of the uranium decay products. By 1985, there were an estimated 191 million tons of tailings at 52 sites – 84 million tons in the state of New Mexico alone. At least half of those sites have been completely abandoned; the companies that owned them are defunct, leaving no one legally responsible for the hazardous wastes. Tens of thousands of people live and work in the vicinity of these piles, which will continue to emit radon for more than 100,000 years. Radioactive dust blowing off these great piles contaminates soil, crops, water supplies, and the people who swallow or inhale the dust or on whose skin it settles. In some areas, windblown tailings have contaminated hundreds of acres of surrounding land; in others, tailings are leaching into the ground water. Many nearby residents are afraid that exposure to the tailings may have caused higher than normal incidences of leukaemia, cancer, and birth defects. No large-scale studies have yet been undertaken to confirm or deny such suspicion.

In 1978 Congress ordered the government to foot the bill for cleaning up the 75 million tons of tailings that had already been generated. As successor to the Atomic Energy Commission, for which most of the tailings were generated, the Department of Energy was made responsible for securing the tailings piles, to standards set by the EPA. The DoE identified 24 sites, containing 25 million tons of tailings, as having top priority for clean-up. The programme was scheduled to be completed by 1990, at a cost that is now estimated at $900 million. The EPA requires that the piles be covered over with earth or concrete so that they are secure against floods and all other natural or man-made disturbance for 1,000 years, or if that is not possible, then for 200 years. However, the tailings will continue to be hazardous for at least 100,000 years.

Environmentalists say that the tailings standards are too lax. The ICRP finds a 1-in-100,000 chance of premature death from radiation exposure acceptable for members of the public. The EPA concedes that the amount of radon leakage allowed under the standards (20 picocuries per second per square metre) would, over a lifetime, give the nearest residents a 1-in-1,000 chance of premature death. This is a higher risk than the EPA has allowed for any other pollutant. In general, the EPA tries to keep exposures to carcinogenic substances to a level that would result in a maximum lifetime risk of from 1-in-100,000 to 1-in-10,000,000. The EPA's original proposal was ten times more strict than the one the agency finally adopted. The laxer limit was chosen, according to Stan Lichtman, who was responsible for developing the agency's tailings standards, because implementing the stricter limit would have been expensive and technically difficult, and because 'there were very few people living near the piles'. Not surprisingly, people who do live near the piles find this argument objectionable. 'Why should those people be less protected just because they don't live in a crowded area?' asks Lynda Taylor, a radiation specialist with the Southwest Research and Information Center in Albuquerque. 'Are their lives less valuable because they don't live in New York?'

One of the first sites to be tackled is in Durango, Colorado, a picturesque Wild West town, perched on the bank of the Animas River. Its one and a half million tons of radioactive tailings sit right in the middle of the tourist district, overlooking the river and the town's charming, old-fashioned railroad station. Twelve thousand people live within three miles of the pile. The DoE, which believes that the pile cannot be made safe in its present location, wants to truck it to a nearby, unpopulated canyon that is a state wildlife preserve, at a cost of more than $200 million. Many citizens who are concerned about the tailings pile where it is are even more worried by the prospect of an operation that would require trucks loaded with radioactive sand to make 272 round trips a day over a two-year period. Some want the DoE to simply lay a blanket of cement over the tailings. Meanwhile, Durango is under something of a cloud. 'Some people have moved here, heard about the pile, and just moved their families out again,' says Greg Hoch, the city planner. Many residents prefer not to dwell on the problem. A large danger sign, with a radiation symbol, used to sit above the pile and dominate the town's skyline; at night the sign was lit up. A few years back, the city fathers had the sign removed, saying it was bad for tourism.

Durango, and most of the other places where there are tailings, also have problems with what the federal government calls 'vicinity properties'. These are buildings contaminated by the tailings. Among the 160 properties in Durango that must be decontaminated are the municipal swimming pool, the garden of the local Holiday Inn, and a liquor shop. Grand Junction, Colorado, population 33,000, has more than 6,200 badly contaminated properties, including homes, shops, offices, churches, and schools. During the 1950s and 60s, the Climax Uranium company allowed people to haul tailings away from its mill for free. Contractors used the tailings to fill foundation holes and to make concrete and brick mortar; gardeners dug the sandy tailings into the heavy local soil; farmers used them as bedding for livestock; the city used them to cushion underground gas lines and sewer lines. The DoE also has to deal with the two million tons of tailings that cover a huge riverside lot in the city's industrial district, with 38,000 people living within a three-mile radius.

The only Eastern tailings site scheduled for decontamination is Canonsburg, Pennsylvania, a town of 11,000, 20 miles from Pittsburg. From 1942 to 1960, a company called Vitro Rare Metals processed uranium in Canonsburg. Like all uranium mills, the US government was its only customer until 1957, when the civil nuclear programme began. Local people had no idea that the tailings presented a hazard; they were grateful to the company for letting them use the sandy wastes in their gardens and their homes. Five years after Vitro Rare Metals left Canonsburg, the Atomic Energy Commission decided that the waste left behind was too radioactive to leave uncovered. In fact, the tailings were still so radioactive that their use and possession required a federal licence. AEC contractors wearing protective suits bulldozed 4,500 tons of the hot tailings into a swampy hole, which was covered over with earth and a porous slag. The contaminated site was then donated to the town, which put a playground there.

In 1980, some residents became disturbed by what seemed like an unusual number of premature deaths in the small town. When a local housewife named Janis Dunn realized that her husband, her brother-in-law, and she herself, had all developed tumours; that a sister-in-law had had breast cancer; and that her husband's mother had died of cancer, she surveyed her neighbours. In 45 households, she found 67 cases of cancer. 'Every house on the three streets nearest the site has at least one case of cancer or a cancer-related death,' she told British journalist Sylvia Collier. Kenny Davis, a retired steelworker, who lives next to a cemetery that borders the company's abandoned land, showed Janis Dunn his family death certificates. Fifteen members of his immediate family who lived near by have died of cancer, including his grandmother, both his parents, three sisters, four uncles, two aunts, and two of their spouses. Mrs Davis has had breast cancer. At least eleven members of the 385-member high school class of 1979 have contracted cancer, and three have died from it.

Also in 1980, tipped off by a health physicist, the *Pittsburg Press* used the Freedom of Information Act to obtain documents showing that the tailings were highly radioactive. The government eventually put

Canonsburg on the top of its list of tailings sites to decontaminate. The job involved cleaning up 160 contaminated buildings and 'encapsulating' 200,000 tons of contaminated material so that it will be secure for 1,000 years. Some homes were so radioactive that they had to be demolished. The work was completed at the end of 1985. More than $20 million was spent to clean up the town, but nothing was set aside to compensate for people whose houses were contaminated or who have developed cancer. The government insists that the tailings have not been a danger to health, and has no plans to study the matter more closely.

Many consumer products contain radioactive materials. In general, radioactive consumer products are exempt from regulation as long as the concentration of radioactive material does not exceed certain levels. The best-known such product is the radium-painted dial. Though most luminous dials are now made with other radionuclides, millions of radium dials are still in use – in watches, clocks, and in the instrument panels of aircraft and ships – and still delivering sizable doses of radiation to their users. Gamma rays easily penetrate the watch or instrument casings, and radon gas and particles of radium paint may also escape the rarely completely leakproof casings and be inhaled or swallowed. Radium pocket watches, usually worn at the hip, deliver about eight millirems per year to the reproductive organs of the wearer. In 1977, there were an estimated 8.4 million radium timepieces in use in the United States, delivering a collective dose of 2,500 person-rems per year. By comparison the total dose to all workers in US uranium mills, enrichment plants, fuel fabrication plants, reprocessing plants, and nuclear waste facilities amounted in 1980 to 2,200 person-rems. Since the mid-1960s, promethium–147 and tritium have replaced radium in most luminous dials. In one sense this change has merely shifted the burden from the dial users to the general public. Promethium and tritium emit beta radiation which, unlike gamma, cannot penetrate glass dial-covers. On the other hand, it takes hundreds of times more promethium or tritium than radium to get the same degree of brightness. Thus, though dial wearers now receive smaller doses, society at large will have more radioactive waste to secure in the future.

Uranium ores are used in some ceramic glazes to produce a shiny orange or yellow colour in crockery and decorative glassware. Such glazes emit gamma and beta rays. Because acidic foods tend to leach the uranium out of the glaze, users of such plates can also receive sizable internal doses of radiation. Fiesta Red, a uranium glaze, was the most popular colour of the ubiquitous Fiestaware line of dishes produced from 1935 to 1971 by a West Virginia china company. Original Fiestaware items now fetch high prices from art deco collectors. The heyday of such glazes was before the Second World War; even nowadays, however, pottery is still from time to time rejected or recalled from department store shelves for being too radioactive. At present the United States government requires no licence for glazes containing less than 20 per cent uranium for dishes, and 10 per cent for glassware, but the US Bureau of Radiological Health has recently recommended that such glazes be banned entirely.

About ten per cent of the enamel used in enamelled jewellery also contains uranium or thorium, according to the US Nuclear Regulatory Commission. One piece – without backing – can deliver as much as 4,000 millirems per year to the skin of the wearer, assuming it is worn an average of ten hours a week. Metal backing could cut the dose to 25 millirems per year. In 1983, the NRC banned the import of uranium-bearing, but not thorium-bearing, cloisonné into the United States, though it could do nothing about the cloisonné already in circulation.

Uranium is also used to give porcelain false teeth the brightness of natural teeth. Although some false teeth are made of plastic, which does not require the use of radioactive materials to produce a natural-looking fluorescence, porcelain false teeth are popular because they are stronger. A survey published in 1980 found that 'no consumer or dental lab technician, and few dentists, were aware of the fact that uranium was a part of the porcelain composition'. On average, porcelain false teeth contain 300 parts per million of uranium, which means that each crown gives a dose to the mouth of approximately 1,000 millirems per year. A full set of porcelain dentures would give a dose to the mouth of 28 rems a year. The likelihood that such exposures would cause gum cancer, a very rare condition, 'simply cannot be estimated with presently known data', according to a report prepared for the NRC. The same report states that 'the fact that most of the benefit a consumer receives from wearing dental porcelain is of a cosmetic nature does make [the acceptability of] any risk associated with this material questionable'. Britain's NRPB has recommended that manufacturers stop this use of uranium altogether.

Thorium and uranium are also often present in the silica and other natural materials from which lenses are made. In 1975, the US Optical Manufacturers Association voluntarily agreed to limit the amount of uranium and thorium in optical lenses, so that a person who wears eyeglasses for 16 hours a day would receive a dose to the cornea of no more than 500 millirems a year. If manufacturers succeed in meeting that limit, then the collective dose to people in the United States who wear glass spectacles would be about 120 person-rems per year, about the same as the collective dose to the country's uranium enrichment workers.

Camping lanterns use thorium to improve the quality of the light they give off. The thorium is contained in a sheath of threads, called a mantle, which is suspended over the flame. When heated, the thorium glows with a bright white light. There is little risk from using a mantle lamp, since the alpha radiation emitted by the thorium is stopped by the lantern's glass cover, but special care should be taken not to inhale or swallow the mantle ash or get it on food while taking out a used mantle. New mantles or those that have not recently been used should be put in a well ventilated place for the first 15 minutes or so of use, to give the radium and other decay products of thorium a chance to burn away.

Many smoke detectors work by using alpha-emitting americium–241 to ionize the air that passes through the detector box. This produces a small electric current which, if interrupted by smoke particles, sets off an alarm. As long as the americium, whose half life is 432 years, stays inside the detector, there is no cause for concern. Disposal of these detectors presents a problem, however. Though each one contains only a small amount of radioactive material, there are many millions of them in the US alone, and the number is expected to grow. In the United States, all used radioactive smoke detectors are supposed to be returned to their manufacturers. The vast majority, however, are probably thrown away with the family garbage, and eventually burned or dumped in a landfill site. Because of their ionizing powers, radioactive materials are also used in some anti-static devices, including lightning rods and record cleaners.

Though the radioactivity of individual items may be negligible, the collective exposure from their manufacture, use, and disposal, may not be. Hospitals, clinical laboratories and physicians need no licence to buy radioimmunoassay kits, a measuring device used in research and diagnosis. Industry sources estimate that a million or more of these kits, each containing from one to ten microcuries of radioactivity, are sold annually in the US alone. 'The suggested way to dispose of the kits is down the drain,' says Dr Gerard Wong, chief of the radiation management section of California's Department of Health.

Probably the most widely used radiation-emitting consumer product – and one that *is* regulated – is the television set. Colour television sets all emit some X-rays. The ICRP recommends a limit of 0.5 millirem per hour measured two inches from the surface of the screen. In 1967 the General Electric Company had to recall 90,000 sets that were emitting excessive amounts of X-rays, prompting the US government to limit emissions to 0.1 millirem per hour at a distance of two inches. The further away from the set one sits, the lower the dose. In the US people who watch TV receive an average dose to the gonads each year of 0.33 millirem for women, and one millirem for men. Men are more at risk than women from TV watching because their bodies provide their reproductive organs with less natural shielding than do women's. Television addicts may like to reflect on the fact that their radiation exposure may be less than that of dedicated readers of glossy magazines. The chemicals used to make paper feel smoother and look whiter contain small amounts of thorium and radium. Exposure to such paper at a distance of 14 inches may entail an exposure of around three billionths of a rem per hour. This, though very low, is still twenty times the dose received by a TV viewer who sits twelve feet away from the screen, according to scientists at West Germany's Karlsruhe nuclear research institute.

As a group, radiation workers receive the brunt of radiation exposure. In theory, they are compensated for the risks of their work. But in one sense radiation workers are not free to accept whatever risks they like. One of the risks of radiation work is genetic damage, which may threaten the genetic health of the population as a whole. Society has always had an interest in minimizing injuries to its members, especially when, as in the case of radiation, the risk extends beyond the exposed individual to future generations.

The radiation workforce is growing rapidly. There are estimated to be from three to five million radiation workers in the world. In the United States, the number of people whose jobs may expose them to radiation has been doubling every 15 years. By 1980, the last year for which records are available, there were one and a half million radiation workers in the United States. As the workforce has grown, so has the collective dose, even though individual doses are declining. From 1960 to 1980 the collective dose for radiation workers rose from 91,000 to 150,000 person-rems.

Even an activity in which individual doses are quite low may have a considerable collective dose, if the number of exposed individuals is large. In 1980, for example, the average dose to American students who were exposed to radiation was 100 millirems. But there were 31,000 such students, so their collective dose was 3,300 person-rems, which is 22 times greater than the collective dose to all nuclear waste workers. Using the ICRP's risk estimate, there is a 40 per cent chance that there will be one death among the group from that year's exposure. If the exposed group doubles or trebles in size, so will the number of deaths, though the risk to the individual stays the same.

The individual can best protect himself or herself by reducing his or her own individual exposure, but society can best protect itself by reducing collective exposure. More efficient training, monitoring, and work procedures and better equipment will accomplish both these goals – lowering the exposures of individual workers and the collective exposure of the workforce. An easier and cheaper way of reducing individual doses is simply to spread the dose among more people – hire more people to do the same job. Unfortunately, this method of lowering individual doses not only fails to lower collective dose, it actually tends to increase the collective dose. One reason for this is that there is inevitably a certain amount of duplication of exposure among workers sharing jobs. In addition, as the workforce grows, so does the difficulty and expense of monitoring, training, and protecting it and the likelihood that safety abuses will occur.

The US nuclear power industry is a case in point. Between 1969 and 1984, the nuclear power industry managed to cut individual doses by an average of 40 per cent. These cuts, however, were largely achieved by simply hiring more people to share the exposure and they resulted in a greater collective dose. In 1984, generating one unit of electricity involved more than 50 per cent more radiation exposure to workers than it had in 1969.

The ICRP has not recommended a limit on collective occupational exposure. Nuclear activists have been campaigning for such a limit since the 1970s. Ralph Nader has proposed a collective dose limit of 500 millirems for every megawatt-year of electricity generated. The average collective exposure per megawatt-year in the United States is 1,500 millirems. In Europe and Canada, where reactors are differently designed, it is 400 and 200 millirems, respectively.

Because there is no limit on collective dose, the nuclear industry has concentrated on spreading exposures to a larger and larger workforce, rather than on the much more expensive alternative of reducing exposures in real terms. Increasingly, the exposure burden is falling on temporary workers, rather than on the permanent workforce. Some nuclear reactor jobs are so 'dirty' in radiation terms, that workers receive their radiation limit in just a few minutes. In 1973, during a shutdown at Consolidated Edison's Indian Point reactor in New York State, it took 1,500 welders, each working for only 15 minutes, to repair six 4½-inch water pipes. Reactor managers rely on temporary workers to do these jobs. These workers, called 'glowboys', 'jumpers', or 'sponges', are laid off once they have reached the dose limit. Some, like the Consolidated Edison welders, are highly specialized workers; others are unskilled labourers whose chief attributes are having not yet received their full allotment of radiation and being small enough to crawl through the 18-inch-wide passageways of reactor vessels to turn a bolt or open a valve.

Most temporary workers are supplied by contractors, who find them through union halls, local employment offices, and newspaper want ads. For skilled workers, the pay is high. If the station is far away, temporary workers also get generous expenses for travel, lodging and meals. Once at the station, jumpers may spend most of their time playing cards or watching television, waiting for the call to work. Two weeks' pay for less than one hour's work is not uncommon. On the other hand the recruits are not covered by health insurance or pension plans, and are often required to sign waivers releasing the contractor and the utility from responsibility for any damages that might stem from radiation exposure.

As reactors get older, the need for temporary workers grows. Between 1972 and 1981 the number of temporary workers at nuclear power plants increased by two thirds. Today almost half of all nuclear power plant workers who are exposed to radiation are temporary employees. In 1984, the Atomic Industrial Forum reported that 'much of the permanent staff [at a typical nuclear reactor], usually 300 to 400 employees, spends a dominant portion of its time planning the inprocessing, work assignments and evaluations of the thousand or more workers that might be brought in during a planned refuelling outage'. Despite hope that some high-exposure jobs will one day be performed by robots, the demand for temporary workers is likely to grow as the decommissioning of the first generation of nuclear reactors begins.

Though there are reports of contractors having rounded up unskilled labourers from college campuses with the appealing slogan, 'A full day's pay for a half day's work', most temporary workers are union members. They are covered by the same regulations concerning radiation exposure, protection, monitoring, and training, as permanent employees are. Despite the pre-job training, which may last from half a day to several days, there is anecdotal evidence that not all workers understand the risks their work entails. Many testify that their instructors told them that their exposure would be kept to a 'safe' level, although there is no such thing as an absolutely safe level of radiation exposure. A 1981 study of temporary workers by Mary Melville of Clark University's Center for Technology, Environment, and Development questions the value of the training. 'We find that the training often reduces the impression of hazard, rather than enhancing it,' the report says.

A number of temporary workers travel from power station to power station, taking jobs as they find them. In 1981, more than 5,264 worked at two or more different stations; 1,138 of them worked at four or more. Forty-six temporary workers received whole-body doses of five rems or more that year. The greatest increase in the temporary workforce has occurred among these 'transient workers', as the NRC calls them. Between 1972 and 1981 their numbers grew by a factor of 40. 'There's a lot of worry about this,' Lauriston Taylor told me. 'There is no way to stop a fellow from getting the full dose at one station and just starting over again at another one.' Though a few utilities are experimenting with a computerized tracking system, most employers rely on job-seekers to tell the truth about their previous exposures, although the truth may disqualify them for a job. 'Among radiation protection supervisory people at the plants there have always been beliefs that workers in some instances do not divulge previous work,' says Melinda Renner of the Atomic Industrial Forum, who has made a study of the subject. Says the NRC's Robert Alexander: 'We don't attempt to protect the worker against himself. If he should be so foolish as to select radiation as the way to take his own life, then he can do that without breaking any regulation.'

CHAPTER TWENTY-ONE

Keeping Track

Keeping track of the amount of radiation the public receives from a multitude of sources is an incredibly complicated job. It involves measuring the amount of radioactive material released into our air and water supplies daily by tens of thousands of radiation users, and then following these radioactive releases as they weave their way through the food chain, from plants to animals and humans. One formidable problem is that different radioactive materials take different paths through the environment. Another is that radioactive materials that emit only alpha rays or beta rays may be impossible to detect once they are absorbed by plants or animals.

Monitoring radiation in the environment and in humans is, in fact, so difficult that it is rarely done. Many people assume that radioactive releases are closely monitored and that government agencies know the radiation exposure of the local soil, water, plants, animals, and people. This is not the case. Some, but by no means all, radiation users are required to continuously monitor their own routine releases for gamma – though rarely for alpha – emissions. Water, soil, and food, however, are checked only sporadically and usually only for gamma radiation. Members of the public are not monitored for radiation exposure because of the expense, the inconvenience, and the fear that such monitoring would cause alarm.

Many smaller radiation users, including many industrial, medical, and research facilities, are virtually unmonitored and unregulated. They are simply assumed to contribute negligible amounts of radiation to the environment. In the United States, for example, the Nuclear Regulatory Commission allows each of the country's approximately 23,000 licensed radiation users to pour five curies of radioactive material down the drain into the public sewage system each year. Licensees are supposed to keep records of how much radioactive material they actually do throw out, but they do not have to report the amount to the government. Officials believe that few licensees go up to the limit. 'If everyone who can do it, does it, that's a problem,' says Hal Peterson of the NRC.

Disposal of some radioactive materials, such as radioimmunoassay kits used in hospitals, is completely unregulated and unmonitored. The government does not monitor the water system closely for radioactive contamination. 'The Environmental Protection Agency does periodically, but not frequently, monitor the water in certain rivers, but it's not enough,' says Peterson. Thus no one knows exactly how much radioactivity is flushed into public waters in the United States each year by many thousands of different users.

Monitoring requirements for large nuclear plants are somewhat stricter than those for smaller operations, though in general such plants are assumed to release as much radiation as they are designed to release. Nuclear power plants in the United States must continuously monitor the amount of radiation in the air at certain points around their plants. Releases of radionuclides into water are usually spot-checked, rather than continuously monitored.

Using these measurements and assumptions about the amount of radiation released into the environment, officials then use mathematical models to estimate how much of that radiation reaches the public. Current estimates are that residents of industrialized countries receive an average dose of ten millirems per year from consumer products, about one millirem from the routine emissions of nuclear power production, and another one millirem from other environmental sources, including nuclear research.

The model is a sort of equation – a description in mathematical terms of the process by which radionuclides move through the environment and expose human beings. The model should take account of all the factors that affect the radiation dose, from the physical, chemical, and biological properties of the radionuclides themselves, to the weather, the type of soil, and local eating habits.

Radiation reaches human beings by many different, sometimes very complicated, pathways. Only a small part of the public's exposure to industrial radiation is thought to come from gamma rays penetrating the heavy shielding that surrounds most radiation sources. Most is from radioisotopes discharged to the air, or to water supplies. Once in the general environment, these radionuclides may be inhaled or swallowed; they may fall on people's skin, on their clothing, or on the ground; or they may enter the food chain, irradiating crops, fish, animals, and the humans who eat them.

Many of these pathways involve mechanisms that are still mysterious to science. Thus, the model-maker has the challenging task of trying to translate complex ecological processes that are not yet fully understood into a mathematical formula. People who use models face the further problem that it may be impossible to get numerical values for many of the factors known to influence radiation dose, such as, for example, the size of the raindrops that wash radionuclides from the atmosphere on to a pasture. For these reasons, even the most sophisticated models are far from precise.

The best-understood pathway is the 'pasture-cow-milk' route by which airborne iodine–131 reaches thyroids. The model expressing that pathway is relatively simple, with only ten variables. Yet the dose estimate it produces may be off by a factor of eleven or more. This is considered 'quite acceptable compared to other uncertainties with which one must deal in the field of public health', says Merril Eisenbud, a member of the NCRP and author of a respected textbook on environmental radioactivity.

Some processes are so complex that it is virtually impossible to translate them into useful models. The model of how plutonium–239 works its way through soil, food, and into human beings, for example, involves so many hard-to-determine variables that the dose estimate it produces may be uncertain by a factor of 100 or more. The movement of radionuclides through water is even less well understood than their movement through the soil, so models of that process yield even less reliable dose estimates. The most tentative models are those that attempt to predict how radioactive wastes stored in rock formations will behave tens of thousands of years into the future.

To make up for this unreliability, model-makers say that they try to err on the side of caution, though some critics say that the models now in use ignore important factors that, if included, would tend to yield higher dose estimates. Scientists at the University of Heidelberg in West Germany have analysed the models that the West German and US governments use in designing their nuclear facilities. Most of the information from which the models were derived came from experiments conducted during the 1950s under the auspices of the US Atomic Energy Commission. The German researchers found that where there was a range of evidence, the model-makers had generally chosen factors 'located at the lower end of the range of realistic values', i.e. those most likely to yield the lowest dose predictions. This picking and choosing, along with the limited amount of experimental evidence available, led to some questionable combinations. For example, the model-makers calculated the movement of radioactive material through the pasture-cow-milk pathway using data on Russian soil, English plants, and American farm animals. The German researchers concluded that the official US and German models seriously underestimate the radiation dose to humans from a number of routine emissions from nuclear facilities, including cobalt–60, strontium–90, iodine–131, caesium–137, and plutonium–239. In some cases, said the Heidelberg team, the model yields dose estimates that are 1,000 times too low.

The best way to check a radiation dose model is to measure its predictions against reality, but this is not often possible. Some models cannot be measured against reality, because they predict what will happen in the future. Other models deal with quantities too small to detect and measure. But even models that could be verified against real measurements seldom are, due to the cost and practical difficulties of doing so. In the words of a 1984 NCRP report, most models have never been 'validated in the field, or evaluated through statistical studies'. Thus, it is impossible to know how much faith to place in them, to know whether they are accurate to within a factor of 10, of 60, of 200, or more, or less – or even whether the model tends to over-estimate or under-estimate the true dose.

On the rare occasions when the predictions of computer models have been checked against real-life measurements, the models have often been wanting. During 1982 and 1983, the US government used specially-equipped Navy planes to track emissions of krypton-85 from its Savannah River reprocessing plant in South Carolina. The model in use, known as the Gaussian model, predicts that airborne emissions from nuclear facilities dissipate fairly rapidly and will be undetectable far away from the plant. Scientists found, however, that the plumes of radioactive air emitted from the Savannah river facility were still intact 200 and even 600 miles away. At 200 miles, the model predicts radioactivity levels of no more than 20 picocuries per cubic metre of air, but the actual readings were more than 5 times higher.

'That standard belief in the industry is that the Gaussian model tends to overpredict, to predict high, but we didn't find that,' says William Lawless, the study's technical director. Other studies have confirmed the validity of the Gaussian model, but most such studies have tested the model only over short distances. In 1984, the Environmental Protection Agency released its own study of Savannah River emissions, which supported the model's predictions, but the EPA took no measurements further than 10 kilometres from the plant. Despite its shortcomings, says Lawless, the Gaussian model 'seems to predict better than the others'. And like other imprecise models, it has been incorporated into regulations intended to protect public health.

Models play a crucial role in the design of nuclear facilities. Engineers use models to forecast the effects of the radioactivity they plan to release. If the models predict that the planned level of emissions will result in unacceptably high doses to the public, adjustments must be made. Sometimes, however, it is the model that is adjusted, rather than the plant design. In the mid-1950s the United Kingdom Atomic Energy Authority chose Dounreay, a village on the north coast of Scotland, as the location for a pilot fast-breeder reactor. Lord Hinton, who as Sir Christopher Hinton was then Britain's chief nuclear engineer, described the planning process years later in a lecture at Strathclyde University. 'We assumed, generously that there would be one per cent leakage from the sphere, and dividing the country around the site into sectors, we counted the number of houses in each sector and calculated the number of inhabitants. To our dismay, this showed that the site did not comply with the safety distances specified by the health physicists. That was easily put right; with the assumption of a 99 per cent containment the site was unsatisfactory, so we assumed, more realistically, a 99.9 per cent containment, and by doing this we established the fact that the site was perfect . . . we knew we had found the right site for the reactor and were quite prepared to adjust what were only guessed figures to support a choice that we knew from experienced judgement was right.'

The ICRP says that emissions from nuclear facilities should not merely meet the public exposure limit, but be as far below it as 'reasonably achievable'. What is reasonably achievable is, of course, not only a function of plant design, but also of operating procedures, which in turn are influenced by cost, convenience, and on at least one occasion by a desire to experiment. The location of the Windscale (renamed Sellafield) reprocessing plant on the west coast of Britain made the Irish Sea a convenient place into which to discharge the plant's radioactive waste. During the United Nations' second 'Atoms for Peace' Conference, in 1958, a Dutch conference participant asked John Dunster, who had designed Windscale's discharge programme, whether, in view of the ICRP's recommendations that any radiation dose be kept as low as possible, the waste was treated to reduce its radioactivity before it was released into the sea. He also asked whether it was 'quite certain that no accumulation of radioactivity can occur from the total release to be expected in the future at great distances from the disposal sites, for instance, by means of a now-unknown mechanism'? Dunster explained that 'in general terms the intention has been to discharge fairly substantial amounts of radioactivity as part of an organized and deliberate scientific experiment . . . One of the principal and, I believe, the most effective methods of carrying out these investigations is indeed to use radioactivity and discharge it and find out what happens to it.'

Since 1952, Windscale has released more than a quarter of a ton of plutonium into the Irish Sea, now believed to be the most radioactively contaminated sea in the world. Seventy-five per cent of the total radiation dose from all the nuclear power stations in the European Economic Community comes from Windscale, according to a British government agency. Radioactive waste from Windscale makes its way all around the coast of Britain. Water currents carry the contaminants into the North Sea and then through the English Channel. Windscale's waste has also been traced to the Baltic Sea, and as far north as Greenland. One of the primary discharges, radioactive caesium, is absorbed by fish in the Irish Sea, which in turn deliver a small but detectable dose to all who eat them – that is, most of the population of the United Kingdom and Ireland. Plutonium, the other main emission, is found in the local soil, air, grazing animals, and housedust, as well as in fish and shellfish. Though the Windscale discharges have been very high – at first for experimental reasons and later because of the volume of waste to be reprocessed – they were always in compliance with the ICRP's public exposure limits, or so Windscale's management has maintained.

Windscale officials did not monitor the local population's exposure to its discharges. Instead they used mathematical models to estimate doses to the environment and the public. These models have turned out to have several shortcomings, Dunster said in an interview in late 1985. For one thing, they under-estimated, by a factor of five, humans' gastro-intestinal uptake of plutonium. In addition, the modellers were told in the early 1950s that local people did not eat shellfish because of the risk from raw sewage. More recent studies have revealed, however, that local people do eat shellfish – and perhaps always have. Planners also overlooked a possible exposure pathway: it appears that rather than remaining forever at the bottom of the sea, bound to the seabed sediments, as planners had hoped, some of the radionuclides, particularly plutonium, return to land. The probable mechanism is that storms churn up the radioactive sediments and bring them to the surface of the sea, where they may be washed ashore and eventually dried out by the sun and blown into the atmosphere. This theory is supported by the detection of plutonium in the air and of plutonium-contaminated sand and dust on nearby beaches and in local homes.

Windscale officials are adamant that, whatever the errors in modelling and monitoring, the exposures caused by the reprocessing plant could not possibly be responsible for the excess of cancers that has been detected in children living near by. James Cutler, a Yorkshire Television producer, discovered in 1983 that the five parishes neighbouring Windscale had a death rate from childhood leukaemia five times higher than expected, and that the death rate from leukaemia for children under ten in the closest parish was ten times higher than expected. A government-appointed inquiry failed to prove or disprove a connection between the leukaemias and Windscale. 'We just can't find enough radiation to have caused all those leukaemias,' Dunster told me in 1985. 'I'm forced to the conclusion that, unlikely as it seems, it's a random cluster.' Less than six months later, the plant's owners admitted that discharges of uranium into the air during the early 1950s were 40 times higher than had been officially recorded. Windscale is now studying ways to cut discharges, says the company, 'to as near zero as possible even though there is no rational cost-effective basis for doing so on risk assessment grounds'.

Estimates of the amount of radiation each facility releases are based at least partly on the predictions of the plant's designers. Because members of the public are not monitored directly, officials can only give an estimate of the average dose that such releases would entail, based on computer models of how radiation travels through the environment to reach humans. The ICRP cautions that 'dose limitation for members of the public is a somewhat theoretical concept, intended for planning purposes, and that it will seldom be possible to ensure that no individual exceeds the dose limit'.

Compared to members of the public and to workers in other industries, radiation workers are monitored closely. Even Robert Alvarez of the

Washington DC-based Environmental Policy Institute, who is generally a critic of the nuclear industry, says, 'You can quibble about the accuracy of their measurement, but this industry has set the trend for the way industries should operate in terms of information-gathering and monitoring.' There are instruments to measure radioactive material in the air, on work surfaces and clothing, and in the blood, urine, sweat or saliva of exposed persons. The devices vary according to the type of radiation likely to be encountered. They include film badges and thermo-luminescent devices for the measurement of whole-body exposure; thimble-like fingertip dosimeters for those who handle radioactive materials; and air samplers for measuring radioactive dust. Even so, there are limits to the completeness and accuracy of the information obtained.

Most monitoring devices do not measure radiation directly, but must be processed and interpreted. The processing and interpretation may be conducted by private companies, government laboratories, or in-house labs. In 1980 an NRC-sponsored study convinced the commission that 'a significant percentage of the personnel dosimetry processors in the US are not performing with an acceptable degree of consistency and accuracy' and that 'the dose received by occupationally exposed personnel may often be considerably different from the dose reported by the dosimetry processor'. As a result, the commission set the first-ever standards for dosimeter processors, requiring them to produce, on average, readings within 50 per cent either way of the actual dose. The standard came into effect in 1988.

Processing quality apart, there are other limitations to what monitoring equipment can achieve. Personal dosimeters, for example, cover only a few inches of the body. Radiation that happens not to hit the film badge or other device will not be recorded. For this reason, workers are sometimes assigned several dosimeters. In 1980 the Southern California Edison Company hired temporary workers to carry out repairs in the steam generator of its trouble-plagued San Onofre nuclear power plant. The workers each wore one film badge on their chests. The chest-level readings 'were well within allowable limits' according to the company, but when, at the NRC's insistence, the company gave workers badges to wear on their heads, it found exposures as high as 7 rems. Most readings were between 3 and 5 rems. Internal exposures are especially hard to measure. If the material inside the body emits gamma rays, the amount inside the exposed person can be determined with the help of a whole-body counter. However, no instrument can detect radioactive material inside the body if it emits only beta or alpha rays.

All employers in the United States are legally required to keep exposure records for any worker likely to receive more than 25 per cent of the allowed radiation dose. For the most part, however, the data is not available to the government or the public; only over-exposures must be reported to the government. The NRC, for example, which licenses 7,500 radiation users, from nuclear power plants to private industry, collects exposure data from only about 500 of them. The other 7,000 do not have to report their exposure data to anyone, even though their workers may receive relatively high doses of radiation. No one knows, for example, what the dose to industrial radiographers really is, since only a few industrial radiography firms are required to report their workers' exposures. NRC staff members, and others in the field of radiation protection, believe that the average exposure of industrial radiographers is around one rem, more than double what federal records indicate. Some experts are also concerned about doses to well-loggers, whose job involves hauling radiation sources such as cobalt–60 or iridium–192 into the field to analyse underground deposits of water or minerals. Working in rough, outdoor conditions, with shielding kept to a minimum so as to keep their equipment lightweight and portable, 'they may get doses as large as radiographers',' says Barbara Brooks, the commission's expert on exposure data.

Data on internal exposure is even more difficult to come by than is external exposure information, because, apart from over-exposures, the only records on internal exposures that the government collects are estimates of uranium miners' exposures. Yet, some occupations entail significant internal doses. 'Studies indicate that for fuel fabricators the internal exposure is at least as great as the external exposure, so you could double the listed exposure for them,' says Brooks. Those most likely to receive internal doses are uranium miners, workers who fabricate and reprocess nuclear fuel for reactors or weapons, and clerks in underground caves where government or corporate files are stored.

In 1981, a survey commissioned by the NRC found that 25 of 46 personnel over-exposures that nuclear power plants reported in the previous five years were caused by errors in monitoring and record-keeping. Another NRC study, conducted in 1980, found that half the power plants inspected had inadequate training and retraining programmes for health-physics staff. One-third of the plants 'had deficiencies in their radiation-protection surveillance procedures and practices', and one-quarter 'had inadequate radiation-protection staffs and serious weaknesses in the control of personnel contamination'.

The job of monitoring the workforce and the workplace, and of keeping radiation exposures as low as possible, is that of the health physicist, a title given to radiation protection specialists during the Manhattan Project in order to obscure the nature of their work. Working in nuclear power plants, in hospitals, in private industry, in universities, and in government agencies, health physicists calculate how long workers can stay in a given area, make sure that they are properly outfitted, and have the power to close down an operation if they feel it will lead to an illegal exposure. Because they must monitor all types and locations of radiation work, health physicists often receive high doses themselves. Today, according to Dr John Poston, of the US Health Physics Society, there are approximately 30,000 health physicists worldwide, about one-third of them in the United States. Many American health physicists received their training in the nuclear Navy. The field is divided almost evenly into professionals, who must have at least a college degree and who generally have done some graduate work as well, and technicians, whose training is mostly on-the-job.

Most health physicists who have attended college have majored in physics, mathematics, chemistry, or engineering; few study biology or medicine. Critics charge that their training tends to produce technocrats who identify with the nuclear industry, rather than independent watchdogs. Many health physicists resent what they feel is the never-ending pressure to lower exposures, and the increasingly complicated administrative procedures proposed to achieve this. Some have come to feel that their greatest problem is not excessive radiation exposure to the workforce, but burdensome government regulations and the irrational fear of radiation. This feeling was clearly expressed in a July 1986 editorial in the *Health Physics Newsletter* commenting on a proposal for stricter exposure standards: 'Occupational exposure involves risks of injury, disease, and death. Life involves risk of injury, disease, and death . . . Any person who selects an occupation that involves risks must be told what those risks are. But, once committed to them with full knowledge, there should be no surprise and recriminations about the consequences.'

CHAPTER TWENTY-TWO

Unnecessary Exposures

More than 90 per cent of all non-background radiation comes to us from medicine. Despite its immense usefulness, medical exposure is now a cause for concern because of its magnitude, because its use is growing so rapidly, and because so much of it is unnecessary. Much of the problem is caused by doctors and dentists who take a dangerously complacent attitude towards medical radiation, fuelled by a potent combination of familiarity and ignorance. In terms of radiation exposure, says Karl Morgan, 'the medical profession is a greater problem than the nuclear industry'.

In the United States, where three out of four people are X-rayed each year, people receive almost half their total radiation exposure from medicine. Recent discoveries that our total radiation exposure is greater than had been realized (owing to high levels of radon in the environment) have reduced the proportion—but not the amount—of radiation that we get from medicine. In 1981 Dr. Arthur Upton, Director of the National Cancer Institute, estimated that radiation from just one year's diagnostic X-rays causes 3,670 deaths and an unknown amount of genetic damage. Upton's estimate is a relatively conservative one. Karl Margan estimates that X-rays cause 16,000 deaths in the United States each year. Spread over the 150 million Americans who had X-rays in 1981, and considering the medical benefits that are presumed to have resulted from those X-rays, even 16,000 deaths may seem like a small number, but many of those deaths and injuries are unnecessary, since specialists estimate that about half of the US medical radiation dose is unnecessary.

The biggest dose of medical radiation comes from diagnostic tests, which entail smaller individual doses than radiotherapy but a much larger collective dose. In 1964 58 million medical X-ray examinations were performed in hospitals in the United States; in 1970 the number had grown to 82 million; by 1980 it was 130 million, a 225 per cent increase in 16 years. The number of X-ray exams conducted outside hospital – 70 million in 1980 – is growing even faster than the number of hospital X-rays, according to Ralph Bunge of the US government's Center for Devices and Radiological Health. Between 1964 and 1980, the average person's annual dose from diagnostic X-rays in the United States increased by 40 per cent.

The picture worldwide is less clear, but even more alarming, according to a survey by the United Nations UNSCEAR committee. Many countries do not collect statistics on medical exposures, and it has generally been presumed that poorer countries conduct fewer radiation exams and thus have lower radiation doses from medicine. UNSCEAR, however, has discovered that in some countries routine medical exposures entail very large doses, both to the patient and the radiographer. In China, for example, 98 per cent of all X-ray diagnosis is done with fluoroscopy, a procedure that requires the patient to stand in the X-ray beam while the physician stands near by studying the image on a screen. A chest X-ray done by fluoroscopy gives a dose roughly 20 times greater than one in which the doctor takes a film and studies it later, but film is expensive in China. It is reported that 90 per cent of the work done by plastic surgeons in China involves repairing radiation injuries.

Another problem in some countries is the quality of X-ray machines. UNSCEAR found, for example, that 60 per cent of the X-ray machines in Bangladesh lacked a control booth: technicians operate them while standing next to the patient, thus getting many doses of radiation every day. Because many machines also lack basic calibration equipment, it is apparently common in some areas to gauge the output of X-ray machines by testing the white blood-cell count of the technicians who operate them. An unusually low count indicates that the machine is emitting too much radiation. This is reminiscent of the early twentieth-century practice of calibrating X-ray machines according to the exposure required to cause a skin-burn. With the world population of 4.4 billion, UNSCEAR estimates that 1.2 billion medical X-rays are conducted each year, 3.5 million dental X-rays, and 22 million diagnostic nuclear medicine procedures.

In addition to conventional X-rays, new forms of radiation are being used to diagnose diseases. Computerized Axial Tomography, CAT for short, rotates hundreds of sharply focused X-ray beams around the body to obtain excellent cross-section images. CAT-scans enable doctors to make much more accurate diagnoses than do ordinary X-rays, and the radiation doses involved may be comparable to those from ordinary X-rays. But the increasing popularity of CAT-scans worries observers who are concerned that physicians and hospitals will be driven to over-prescribe the scans in order to pay for their expensive CAT equipment.

Another growth area is the use of radioactive chemicals, called radiopharmaceuticals, to diagnose and treat diseases inside the body. At one time, radiopharmaceuticals were considered inappropriate for anyone under the age of 18, and their labels carried a warning to that effect, but they are now used increasingly often to diagnose children's ailments. The use of radiopharmaceuticals increased by five times between 1960 and 1970 and has grown even faster since then. In 1980, in the United States alone, radiopharmaceuticals were used approximately 11 million times, mostly for diagnosis.

Each medical procedure entails a different dose of radiation. A chest X-ray taken using modern equipment may give a skin dose as low as two millirads (for X-rays a millirad which measures the amount of radiation absorbed is virtually equal to a millirem, which measures the biological effect of that radiation). In comparison, an examination of the lower gastro-intestinal tract, which involves a combination of fluoroscopy and ordinary X-ray photographs, can easily deliver a skin dose of more than 1,000 millirads. A complete abdominal exam can give the patient a skin dose of 5,000 millirads. (It is usual to compare X-rays in terms of skin doses, even though the skin may not be the most vulnerable organ exposed, because the most vulnerable organ will vary from procedure to procedure.)

The 1983 US government statistics on X-ray exposure, the most recent available, also show extreme variations in dose for identical X-ray exams. The median dose to the skin from a chest X-ray is 15 millirads, but some patients receive as little as 2 or as much as 253 millirads. The skin dose for an abdominal exam, one of the most common types of X-ray, ranges from 21 to 2,575 millirads. Skin doses from a dental bitewing (an X-ray of several upper and lower teeth, so called after the wing-shaped film the patient bites on) vary from 52 to 2,389 millirads. The wide range reflects differences between machines, as well as between machine operators. A 1979 nationwide survey by the US government found that 30 per cent of all X-ray machines failed to comply with one or more federal tests.

The variations in patient exposures from state to state, office to office, and machine to machine, is alarming and at the same time reassuring, since it implies that the worst cases are capable of substantial improvement. The State of Illinois has had impressive success in lowering X-ray exposures since the early 1970s when its department of public health began a campaign to encourage doctors to improve their X-ray techniques. Between 1972 and 1979, average exposures in Illinois fell by 49 per cent for dental X-rays, by 30 per cent for abdominal X-rays, and by 27 per cent for X-rays of the lumbar spine. The combination of government monitoring and better equipment lowered the average skin dose for a single dental X-ray in the US from 1,140 millirads in 1964, to 910 in 1970, and 400 in 1981.

The story for medical X-rays is less encouraging overall. Though many establishments have succeeded in cutting their exposure rates, the most recent comprehensive government survey, covering the years 1974 to 1981, found that 'mean exposure levels for medical examinations remained relatively stable during the six-year period in spite of wide variations that occurred among facilities'. The mean dose to internal organs from chest X-rays actually increased by an average of 35 per cent between 1974 and 1981.

Testifying before a congressional committee in 1979, Daniel Donohue, then president of the American Society of Radiologic Technicians, estimated that 40 to 50 per cent of all X-ray technicians 'have no recognized credentials'. John Villforth, director of the Center for Devices and Radiological Health, told the same committee that 20 per cent of all dental X-rays are taken by a completely uncredentialled person. Another witness, Eleanor Walters of the Environmental Policy Institute, pointed out to legislators that in Vermont a landscape gardener who X-rays tree trunks must be licensed, but a hospital technician who X-rays patients need not be.

California has the nation's most stringent X-ray regulations and most thorough monitoring programme. No licence is required of dentists, but California is the only state that requires physicians who use or supervise the use of X-ray equipment to be licensed, and it is one of only 15 states that require X-ray technicians to be licensed. In the other 35 states, anyone can operate an X-ray machine – absolutely no training is required. In many doctors' and dentists' offices, secretaries or office clerks administer X-rays as part of their normal duties.

Neither their years of training, nor the state's strict licensing requirements, ensure that technicians, dentists, and doctors in California really understand the radiation doses they administer to patients. I spoke to one licensed X-ray technician in San Francisco, a recent graduate of an American Dental Association-accredited dental assistant's course, who did not know in what units radiation is measured. Neither did her employer, a recent graduate from Marquette University's School of Dentistry, who kindly demonstrated her X-ray equipment for me. 'I think I remember hearing somewhere that it's the same as spending a day in the sun,' she replied, hesitantly, to a question of mine, 'but I don't know if that's right. I could try and find out for you. I know we had that in school – you know, all those technical things that you forget as soon as you learn them.'

'This is not atypical, unfortunately,' said Constance Bramer, a radiation protection specialist for the California Department of Health Services, referring to my informant's uncertainty about radiation doses. 'Dentists should be far more knowledgeable; they should know their own field; they really should know the dosage.' In the late 1970s, state inspectors found that 51 per cent of California's dental practices exposed their patients to radiation doses described by Donald Bunn, senior house physicist in the California State Department of Health, as 'over the acceptable range'.

Many physicians, too, have trouble qualifying for a California X-ray licence. 'A good percentage of the doctors who take our test fail the first time,' remarked Bramer. They get one more chance, but even so only about 80 per cent of doctors eventually pass the exam, according to Joseph Ward, the former head of the department's radiological health section. The most recent survey of X-ray equipment and techniques, in 1978, found that 71 per cent of California's radiologists and 75 per cent of its hospitals had unsafe equipment, and that 17 per cent of the hospitals and 26 per cent of the radiologists used unsafe procedures. 'I would guess that things have improved since then,' says Bunn, 'because we've been able to hire more staff as a result of that survey.' The state now has 30 X-ray inspectors, who are responsible for checking the 48,000 X-ray machines in the state and advising operators on how to lower exposures.

California's guidelines as to what exposures are acceptable vary according to the age of the X-ray machine being used. Many pre-1970 machines operate on 50 kilovolts: a single dental X-ray from those machines is allowed to deliver a skin dose of up to 475 millirems. The same procedure carried out by a 100 kilovolt machine should expose the patient to no more than 150 millirems. The state does not require that doctors use the most efficient machines available because it does not want to, as Bramer put it, 'dictate to medicine'.

Many doctors and technicians wrongly believe that long exposures are essential to obtaining useful X-rays. In fact excellent images can be obtained on relatively short exposures, if the film is properly developed. Unfortunately, says Bunn, 'there's not much training in the schools regarding this subject – at least in the past. They very rarely change their settings and this is where the problem lies; they don't know how to optimize the exposure. In general, dentists don't know how to develop X-ray film'. Bramer agrees: 'Developing properly is really the critical thing.' Using ancillary equipment, such as filters, beam restrictors, fast film, and rare-earth screens, can also help lower exposures. Image-intensifying screens, which use the class of chemicals called rare earths, can cut exposure times in half. Though relatively new on the market, says Bunn, such screens are rapidly gaining acceptance among doctors. Not all new equipment allows exposure times to be reduced, however. A device called a grid, which improves images by absorbing X-ray scatter, requires exposures to be increased by almost a factor of three. Using a grid can easily cancel out any savings in exposures made by fast film, but it does produce sharper images, which should make diagnosis easier.

A common, and easily corrected, cause of over-exposure is a radiation beam that is too large for the X-ray film. Any radiation that falls outside the film is wasted – an unnecessary hazard. In California, dentists are required to restrict the beam diameter to 2.75 inches, which covers an area of 5.94 square inches. But most dental film has an area of only 2.19 square inches, so almost two-thirds of the radiation exposure is unnecessary. This means that every dental X-ray given in California, even by the most efficient technicians using the most advanced equipment, delivers at least three times as much radiation to the patient as necessary. The beam-to-film ratio for most other types of X-ray exam is better, but still wasteful. The average chest X-ray, for instance, irradiates an area 20 per cent bigger than the film being exposed.

In 1975 the Bureau of Radiological Health (now the Center for Devices and Radiological Health), tested operators and found that 63 per cent of those without credentials did not properly restrict the X-ray beam. Forty per cent of the credentialled operators failed the same test. The situation was much the same nine years later when the bureau published the results of a survey of X-ray trends between 1974 and 1981. Non-credentialled operators who do not work in hospitals, where there are qualified supervisors, generally use larger beams than credentialled operators performing identical exams, the survey showed.

Another way to reduce exposures is to use lead shields, especially for the reproductive organs. Women's reproductive organs are more vulnerable to radiation-induced damage because from birth the ovaries contain all the eggs a woman will ever have. Thus X-raying a woman who is not pregnant could damage her future children. It is sometimes impossible to shield a women's ovaries from the X-ray beam; men's testicles, however, can always be shielded. The Center for Devices and Radiological Health estimates that shielding could reduce the radiation dose to male sex organs by 75 per cent. There are few legal requirements for shielding, apart from California's rule that dentists must use lead aprons during all X-rays. In the United States as a whole, according to the American Dental Association, only one dentist in four shields patients during X-rays.

There is no way for a patient being X-rayed to know what dose he or she is receiving. Nor does the doctor know. The dose depends upon the machine being used and the way it is used – the length of the exposure, the size of the beam, and the positioning of the film. All a patient can know is that if the doctor or technician is well trained, and uses modern, well maintained equipment and the best procedures, the dose is likely to be within the acceptable range for that procedure.

Besides being exposed to unnecessarily high doses of radiation from each exam, patients must often endure multiple exposures because the operator fails to obtain a useful image on the first try. On the basis of several studies, the Center for Devices and Radiological Health estimates that more than one in ten X-ray exams are worthless and must be repeated. Moreover, many exams that are not repeated are of such poor quality that they are unusable, and thus constitute wasted exposure to the patient. In 1975 a leading chest radiographer, Dr Benjamin Felson, said that half the films used to diagnose black lung disease among miners were unreadable.

Even the lowest exposure is unacceptable if it is unnecessary, and according to the Center for Devices and Radiological Health one in five X-ray exams *is* unnecessary. There are several incentives and pressures to order unnecessary exams. What Lauriston Taylor has called 'improper purposes' for X-rays include impressing patients; showing insurance companies that the work has been done; obtaining evidence in case of malpractice suits; and justifying investment in X-ray equipment.

'X rays are commonly used to treat anxiety in patients or in doctors,' according to an editorial in the journal *Radiology*. Many patients consider X-rays an indispensable element of the medical armoury and feel cheated if they are not X-rayed. In addition, insurance companies often insist upon X-rays to prove that work was needed or has been carried out.

The growing number of malpractice suits also encourages the prescription of unnecessary X-rays, both to satisfy demanding patients and as evidence for the doctor. In the book, *A Practical Medico-Legal Guide for the Physician*, doctors are told: 'An excessive number of X-rays may be taken, but they are necessary . . . to protect the physician from a malpractice claim. A lay judge or jury often considers the examination incomplete without pertinent X-rays. The patients themselves frequently share this opinion, often attaching unwarranted importance to radiological examinations.' Seventy-five per cent of the doctors polled by the American Medical Association in 1977 said they ordered extra X-rays for their own legal protection. Radiologist John McClenahan gave an example of defensive X-raying: 'If a tennis player suffers elbow pain after a truck scratched the fender of his car, a radiologist will be called on to take pictures not only of the elbow, but of a shoulder . . . a forearm, a neck, chest, and, after the diarrhoea ensuing as the result of stress imposed by the accident, of the patient's entire gastro-intestinal tract.'

Another inducement for taking unnecessary X-rays is the charge doctors and hospitals can make for them. 'You buy a piece of equipment and then you've got to do enough business to write it off,' explains James English, a radiologist. Evidence to a 1979 congressional committee investigating unnecessary X-ray exposure indicated that patients of non-radiologist doctors who have their own X-ray equipment are twice as likely to be X-rayed as patients whose doctors do not perform their own X-rays. X-rays are also a major source of income for hospitals, though a recent change in the way Medicare funds are dispersed has reduced the incentive for ordering unnecessary X-rays. The Center for Devices and Radiological Health estimates that each year hospitals in the United States order more than 18 million unnecessary X-rays. The legal question of who owns X-ray films has not been finally decided. If patients could give copies of past X-rays to new doctors and dentists, there would be less need to order new ones.

Although the mass screening programmes for tuberculosis that were popular throughout the 1950s and 60s have mostly ended, similar programmes in which large numbers of people are routinely X-rayed still thrive. The greatest potential for savings in exposure today is probably in these routine screening programmes.

Many hospitals still insist that all patients have chest X-rays, although expert panels have repeatedly recommended an end to the practice. In 1979 the Joint Committee on Accreditation of Hospitals recommended an end to routine X-raying of patients entering hospitals. The next year an expert panel convened by the Department of Health and Human Services said that routine chest X-ray programmes did not yield enough useful information to justify their cost and extra radiation risk. Not until 1981 did the federal government drop its requirement for regular chest X-rays for 160,000 government employees. Even so, US hospitals administered 35 million routine chest X-rays in 1986.

X-rays of the female breast, called mammograms, can reveal the early stages of breast cancer, the leading cause of cancer death in American women. This technique, known since the 1930s, gained in popularity during the 1960s and 70s, after Happy Rockefeller and Betty Ford underwent mastectomies, making millions of women aware of the risk of breast cancer. In 1973 the American Cancer Society, in co-operation with the federal government's National Cancer Institute, established 27 centres where women 35 years and older were given mammograms, free of charge, and were instructed in self-examination for breast cancer. In 1974 there was almost a stampede for mammograms. In some centres, according to Rose Kushner, the director of the Breast Cancer Advisory Center, 'waiting lists for mammograms were often months long'. Within three years, several hundred thousand women were examined and 1,800 cases of breast cancer were detected.

What was largely ignored in the enthusiasm for mammograms is that mammograms can cause, as well as identify, cancer. The breast is extremely sensitive to radiation-induced cancer, and mammograms deliver high doses of radiation compared to most diagnostic operations. It was estimated that this mammography programme produced five cancers for every one detected. A number of doctors began questioning the wisdom of the push for mammograms. Studies indicating that mammography did not increase the survival rates of women younger than 50 caused the American Cancer Society to recommend in 1976 that women younger than 50 should not be mammogrammed unless they have symptoms or a family history of breast cancer.

In September 1987, however, the American Cancer Society changed its guidelines again. It now advises all women to have a 'baseline' mammogram at age 35, followed by a mammogram every other year from age 40 to 50, and one annually from then on. The reason for the change is that the odds have improved for detecting breast cancer without inducing it, because new equipment and techniques make it possible to keep the skin dose under 400 millirads per film. Earlier exposures were generally much higher. The American Cancer Society warns patients to make sure that the person giving the mammogram is specially trained in mammography (at present there is no formal certification, but the American College of Radiology is instituting a programme); that the equipment used is designed especially for mammography; and that it is inspected and calibrated annually by the state department of health.

Unfortunately many mammography clinics still fall far short of these ideals. During a recent seminar at the University of California at Berkeley, John Gofman warned women seeking a mammography that 'you are very likely to walk into a place that doesn't know what dose it is using and doesn't measure its dose, and you may get ten times the recommended dose'. Jacob Fabrikant, an opponent of Gofman's in the nuclear debate, added, 'Rarely in front of a microphone would I go on public record saying John Gofman is right. But he is certainly right in this. There are horrible abuses.'

The realization that a foetus exposed to even a single abdominal X-ray has an increased chance of getting leukaemia (and, it has recently been discovered, of being mentally retarded), has led to attempts to ban all non-emergency abdominal X-rays of women of childbearing age outside the first ten days of the menstrual period. The medical profession has resisted that suggestion. A 1976 telephone survey by the New York Public Interest Research Group found that many doctors failed to ask women whether they might be pregnant before giving them an abdominal X-ray. Instead of requiring doctors to ask a woman if she is pregnant before X-raying her, the US government and various medical associations launched a poster campaign a few years ago, urging pregnant women to remind their doctors not to X-ray them. In 1980, the year in which major American medical groups recommended that doctors not routinely X-ray pregnant women to check the size of their birth canal, 266,000 pregnant women in the US were X-rayed for that reason.

Employers sometimes offer – as an attractive job benefit – free annual physicals with high dose X-rays of the upper and lower gastro-intestinal tracts, kidney, and spine. More often they require prospective employees to undergo X-rays for legal reasons. The most common such exam is the lower back (lumbar spine) X-ray as a pre-requisite for jobs involving heavy lifting. Lumbar spine X-rays give higher doses than almost any other common X-ray, and there is no clear evidence that they actually help employers to identify which would-be workers are likely to injure themselves by lifting heavy weights.

One of the most over-used X-ray procedures is the skull X-ray. A full skull exam requires at least four films, and entails an average whole-body dose of roughly 35 millirems. Studies have shown that certain combinations of symptoms are almost entirely reliable indications of skull fracture, and that X-rays add nothing useful to the information available in a physical inspection. Even if X-rays do show a skull fracture where none was suspected, this finding does not in most cases alter the medical treatment. Nonetheless many physicians insist that anyone with head injuries have a skull X-ray, to determine if there is a fracture. The US government estimates that there are 800,000 unnecessary emergency room skull X-rays every year. In 1978, the cost of unnecessary skull X-rays was estimated to be more than $31 million per year.

Another common occasion for unnecessary X-rays is the routine dental X-ray. Many dentists automatically subject all patients to a full-mouth set of X-rays, which involves as many as 18 separate films, on their first visit and every 12 or even 6 months thereafter. The average whole-body dose from a full-mouth exam is about 20 millirems. The American Dental Association has said that a full set of X-rays should not be needed more than once every five years, but a 1981 meeting of experts on dental radiology concluded that X-rays should not be given according to a timetable at all. They should be given only if a visual examination of the patient indicates a clear need for more information. In the keynote address to the conference, Lauriston Taylor said, 'On its face this would seem to exclude routine full-mouth X-ray examination, even for new patients. It is to be noted that dental schools differ sharply on this practice.'

Since they will drop out within a few years anyway, baby teeth should not as a rule be X-rayed. Karl Morgan offers the following advice regarding dental X-rays for young children: 'If your dentist says, "I think we better X-ray your son's teeth," tell him, "No, I don't want them X-rayed." If the dentist can explain why and give a good reason for it, you might want to consider it, but also you might want to consider going to another dentist.'

CHAPTER TWENTY-THREE

Out of Control

In theory, the ICRP's rule that members of the public must not be exposed to more than 100 millirems of radiation a year applies to all exposures whether planned or accidental. In practice, however, accidental exposures are as exempt from regulation and control as natural background and medical exposures. It is impossible to say what the global or average radiation dose from radiation accidents is because there is no central registry for such accidental exposures and no authoritative source on the total amount of radioactivity they have released into the environment, or the amount that has reached human beings. In addition to well-publicized disasters such as Chernobyl or Three Mile Island, there are hundreds of smaller nuclear accidents every year.

If they are announced at all, these accidents are followed by a reassuring statement from local authorities that the radiation released was well within 'internationally accepted safety limits' or words to that effect. This is misleading in the sense that there is no such thing as an absolutely safe level of radiation exposure. The true sense of such statements is that the risk posed by the exposure has been deemed by the ICRP or by local officials to be an acceptable risk. Rarely, however, do authorities know exactly how much radiation was released, what paths it travelled, and what the dose to the public was. Given the difficulty of tracking even planned releases of radiation through the complex pathways by which they reach human beings, it is not surprising that most exposures from radiation accidents can only be estimated.

The world's worst nuclear accident, in terms of uncontrolled radio-activity released, has been the explosion of almost 500 atomic bombs above ground. Atmospheric testing has added to the world's environment approximately 27 billion curies of iodine–131, 34 million curies of caesium–137, 22 million curies of strontium–90, and 350,000 curies of plutonium–239 among other long-lived fission products. The estimated dose to each person on earth from this fallout is now four to five millirems per year. The United Nations calculates that the collective dose from the fallout that will reach earth by the end of this century will be 540 million person-rems. According to ICRP risk estimates this amount of radiation will cause 67,500 fatal cancers and more than 43,000 genetic defects.

Nuclear testing has now largely gone underground, though some countries, including China, India, France and South Africa, have since exploded bombs above ground. In underground testing, the same fission products are released, but instead of being released to the air, they are mostly trapped in the earth. According to Merril Eisenbud, 'the quantities of debris that remain underground after such tests are huge, but objective evaluation of possible long-range off-site risks has not been possible because few of the basic data have been made available'. The United States has performed more than 500 underground tests at the Nevada Test Site since 1963, some of which were conducted jointly with Britain. The US currently conducts about 20 tests a year. The Soviet Union has conducted closer to 400 underground tests since 1963. France, which moved its testing underground in 1975, has exploded more than 100 nuclear weapons on or under Moruroa atoll in French Polynesia. In the United States alone, underground testing has produced approximately 500 million tons of fission products.

Underground tests are generally conducted in enormous tunnels or shafts from one-eighth of a mile to a mile deep. The weapon and a giant canister full of sensitive detection equipment are set in the hole, which is then filled up with earth. As the weapon is detonated, the ground often collapses, creating a deep crater. The blast sends shock waves out for miles. Because the Nevada Test Site is near two fault systems, underground blasts there have triggered hundreds of earthquakes in the past 25 years.

The United States suspends underground tests if the wind is blowing towards Las Vegas, because of the possibility of radioactive gas escaping to the atmosphere, a process known as 'venting'. The government does not announce all nuclear weapons tests, nor does it say if a test has vented or not. In 1979, however, US officials told a congressional hearing that about one in ten underground tests had vented radioactivity beyond the test site borders. The most serious venting occurred during the Baneberry test in 1970, when huge plumes of radioactive smoke rose over the desert. The fallout cloud drifted east and north, depositing fallout across the United States and into Canada. Four workers contaminated by Baneberry died soon thereafter of leukaemia. The most recent test to go badly wrong was Mighty Oak, in April 1986, when the DoE was forced to deliberately release radioactivity from the test shaft to the atmosphere, using a series of filters to reduce environmental contamination.

Preston Truman, director of Downwinders, an organization that is pressing for a comprehensive test ban, fears that the geological stability of the earthquake-prone test site has been so undermined by hundreds of tunnels and craters that further testing could cause a monstrous ground failure, which might allow thousands of tons of pent-up radioactivity to escape to the atmosphere. 'Mount Rainier [the local name for one of the Nevada site's test grounds] is a Challenger waiting to happen,' says Truman, referring to the 1986 space shuttle disaster. 'They have already blown it apart so much that one of these days there is going to be a massive release of radiation.'

The explosion and fire at the Chernobyl power reactor produced the greatest single release of radioactivity in history. In ten days during the spring of 1986, Chernobyl spewed at least 36 million curies of radioactivity across the world, including more than 7,000,000 curies of iodine–131, 1,000,000 curies of caesium–137, 220,000 curies of strontium–90, and 700 curies of plutonium–239. Parts of Europe and the Soviet Union were more contaminated by Chernobyl than by all the nuclear weapons tests put together. The aftermath of the Chernobyl disaster is well known: the radiation alerts in virtually every European country, millions of people drinking iodine solution in hopes of preventing their body from absorbing radioactive iodine; children being kept indoors; officials in many areas banning the sale of milk, vegetables, and meat. In Russia, the 135,000 people who were evacuated from their homes within 18 miles of the reactor may not be allowed to return for several years. The soil around Chernobyl, which is in one of the Soviet Union's prime agricultural areas, will not be suitable for farming for at least ten years. Thirty-one people died shortly after the fire. Estimates of the number of fatal cancers the fallout will cause over the next few decades range from several thousand to more than 100,000 – half in Russia, half in other parts of Europe, and a small number in Asia and the United States. According to Jay Gould, radiation from Chernobyl weakened the immune systems and caused the premature deaths of from 35,000 to 40,000 people in the United States alone during the summer after the explosion.

One reason for the widely divergent estimates, apart from disagreements about how hazardous radiation is and how it attacks the human body, are the gaps in the monitoring system that make it difficult or impossible to predict the internal dose from contaminated food. After Chernobyl, the Environmental Protection Agency admitted that its monitors were incapable of detecting radioactive iodine that is in the form of a gas. As a result, the EPA's estimates of radiation levels were about three times too low. Radiation protection specialists in Finland, where a more efficient type of filter is used, said that the filters used by the EPA and other countries detect as little as 15 per cent of the radioiodine present in the atmosphere.

One result of inadequate monitoring was that many areas of Europe were not recognized as having been contaminated by Chernobyl until weeks or months after the explosion, and restrictions on selling radioactive produce were introduced late or not at all. Lack of information may be a boon to officials who are loath to risk frightening the public. In Britain, for example, ministers announced that pollution from Chernobyl would disappear in a few days or weeks, and that no restrictions on the sale of meat or milk would be necessary. This attempt to establish calm backfired seven weeks later, when the government changed its mind and banned the sale of sheep over a wide area of northern England, Wales, and Scotland due to dangerously high levels of radioactivity, thoroughly alarming the tens of thousands of people who had in the meantime eaten lamb from those very farms.

Still shrouded in mystery are the effects of another Soviet nuclear disaster, an explosion that occurred in late 1957 or early 1958 among nuclear wastes stored at a military site in the Ural Mountains. Although hints of the disaster, which apparently killed hundreds of people and contaminated thousands of square miles, appeared in the Soviet medical literature, it was largely ignored in the Western press until 1976, when the exiled Russian biochemist Zhores Medvedev, assuming the accident was well known in the West, mentioned it in an article for the British magazine, *New Scientist*. To Medvedev's surprise his reference to a catastrophic disaster in the Urals was met with anger and denial from nuclear experts in Britain and elsewhere, who accused him of trying to influence the debate about the safety of nuclear power. This conflict caused Medvedev and other researchers to search for evidence of the alleged explosion. They found more than 100 articles about the ecological effects of large-scale radiation contamination in Soviet technical journals of the 1960s and 70s. The contaminated area contained between 100,000 and 1,000,000 curies of strontium–90. Later, a Freedom of Information Act lawsuit forced the CIA to admit that it had evidence of a nuclear disaster in the region.

Perhaps the most chilling discovery came from investigators who compared old and new maps of the area. The newer maps omitted the names of scores of villages and towns in the region where the disaster occurred. An area of hundreds of square miles had been deserted; the people living there forced to relocate. Lev Tumerman, another *émigré* Russian scientist, confirmed this with his own account of driving through the region in 1960. Tumerman said that road signs 'warned drivers not to stop for the next 30 kilometres and to drive through at maximum speed. On both sides of the road as far as one could see the land was "dead"; no villages, no towns, only the chimneys of destroyed houses, no cultivated fields or pastures, no herds, no people . . . nothing.' Soviet officials have still not admitted the catastrophe.

Prior to the Chernobyl explosion, the most notorious nuclear accident was the one at the Three Mile Island reactor in the spring of 1979. Official records indicate that 15 curies of radioactive iodine was released from Three Mile Island, resulting in an average dose to people living within 50 miles of the reactor of about one millirem – well within established exposure limits. As Merril Eisenbud points out, the dose 'was not measured – this is not possible even with the best available instrumentation – but was estimated by using environmental dosimetric models'. Critics, who charge that the monitoring system was inadequate and that the models used underestimate actual exposures, say that releases and exposures may have been hundreds of times higher.

Less than four months after the accident at Three Mile Island, another accident occurred at Church Rock, New Mexico. Nuclear officials called it 'the worst incident of radiation contamination in the history of the United States', but because it happened in a relatively isolated part of the United States and affected a dispersed population, most of whom are Navajo Indians, it received far less attention than did the accident at Three Mile Island.

The Church Rock disaster began on 16 July 1979, when an earthen dam holding radioactive uranium tailings burst, releasing 93 million gallons of liquid wastes and more than 1,000 tons of solid wastes into the Little Puerco River, an important source of water for washing and for watering livestock in that semi-desert region. United Nuclear Corporation, the owner of the dam, eventually shovelled 3,500 tons of contaminated sediments out of the riverbed, but that was not enough to decontaminate the river water, which remained unfit for animals or humans. Officials posted signs warning people not to drink from the Little Puerco River, but, as Navajo herdsmen pointed out, their free-ranging animals can't read. Researchers from the federal government's Center for Disease Control found elevated radiation levels in the tissue of some grazing animals, and issued a warning against eating the livers or kidneys of local livestock. Eventually some of the contaminants seeped into the earth: tests found high levels of radioactivity and heavy metals in groundwater 30 feet and more below the surface.

The Little Puerco is a tributary of the Little Colorado, which drains into the Colorado River, which in turn feeds Lake Mead, from which Los Angeles takes some of its drinking water. The water from the Puerco is so diluted by the time it reaches Lake Mead, say government officials, that it is safe to drink. But the Church Rock disaster is not the only time that tailings have entered the river system that leads to Lake Mead. Between 1959 and 1977 there were, according to the NRC, at least 15 accidental tailings releases: seven dam failures, six pipeline accidents, and two floods. In addition, radioactive contaminants from undisturbed tailings piles are easily washed by rainwater into streams and underground water supplies. There are about 190 million tons of tailings in the western states releasing radon to the air, and radioactive materials and heavy metals to the groundwater. In the early 1960s, bottom sediments in Lake Mead were three times as contaminated with radium as were sediments from rivers upstream of the uranium mills. Neither the Colorado River nor Lake Mead is now monitored for radioactivity. Several state and federal water quality officials have said that they would be interested in the results of such monitoring, but that there is no money for it.

The United States' second-largest accidental release of stored radioactive waste occurred in 1973 at the Hanford nuclear complex, where high-level liquid waste is stored in more than 150 large underground tanks. Officials first noticed on 9 June that some of the waste was missing from tank T–106, the oldest and largest of the tanks, with a capacity of more than half a million gallons. No one was certain how much waste had seeped out, nor how much had been there in the first place, but they started pumping out what remained in T–106 anyway. Not until the following day did the authorities realize how serious the problem was. From the incomplete records available, they calculated that the leak had begun 'on or about' 20 April. Each day for almost two months 2,500 gallons had poured out of the tank and into the sandy Hanford soil. Altogether 115,000 gallons, containing 40,000 curies of caesium–137, 14,000 curies of strontium–90, 4 curies of plutonium, and various other radioisotopes, had leaked out. When other leaks from the underground tank farm are included, Hanford's losses total at least 500,000 gallons of radioactive liquid since 1958. Hanford officials have also poured another 200 billion gallons of dilute radioactive and other hazardous liquids out on to the ground since the 1940s. However, these emissions were not accidental but part of Hanford's routine waste disposal practice. The leaks caused much public alarm when they were made public during the 1970s, but DoE officials maintain that even if the wastes eventually reach the Columbia River, they will not raise radioactivity to hazardous levels.

'Material Unaccounted For' is the name the Department of Energy gives to lost nuclear fuel. At present more than four tons of plutonium and uranium are unaccounted for. DoE investigators deny that the missing materials have been stolen or inadvertently released to the environment; they say that they are probably stuck in pipes or filters, or that they only *seem* to be missing as a result of poor record-keeping. An investigation by the US Government Accounting Office of the Materials Production Center, a DoE plant in Fernald, Ohio where uranium is processed into nuclear fuel, found that the contractor had records of releasing 96 tons of uranium dust into the air and another 74 tons of uranium into local waters, but an additional 337 tons of uranium had somehow left the plant without being accounted for.

Abandoned radiation sources have caused severe radiation accidents on several occasions. The worst such case began on 13 September 1987 in Brazil, when a scrap metal dealer bought part of a disused radiotherapy unit from some teenagers who had found it in an abandoned radiation clinic. The dealer, Ivo Ferreira, smashed open the metal cylinder, releasing 1,400 curies of powdered caesium–137. Fascinated by the pretty blue powder, which glowed in the dark, Ferreira gave some to his wife, his daughter, and several friends. Several people, including young children, rubbed it on their bodies. When Ferreira's family and friends grew increasingly ill with nausea, headaches and fever – the early signs of radiation poisoning – his mother-in-law blamed the powder, saying that it was cursed. Ferreira carried the caesium to local health officials, leaving a trail of contamination on his journey by public bus. Three people died within a few weeks; another has since followed. Seventeen people have severe radiation poisoning. According to Brazilian officials, the number of people contaminated is between 250 and 1,000. Outside experts say that several thousand people may have been exposed. The pathways of exposure are bizarre and impossible to predict. Some of the powder was wrapped in a piece of paper, which was later thrown away. The paper was found to have been recycled into lavatory paper which was widely sold. Twenty-five homes have been abandoned, and an area of 2,500 square metres in the centre of the city has been cordoned off for decontamination, which the authorities say may take more than a year.

In a similar case, four years earlier, a therapy unit containing 400 curies of cobalt–60 was tossed into a junkyard in Mexico. The cobalt pellets were taken with other scrap metal to two foundries, which recycled it into various items, some of which were shipped to the United States. Several months later, radiation was detected in some steel table legs, and the source traced back to the discarded therapy unit. No one has yet died of the exposure, but one man has been rendered sterile as a result, and some cancers are likely to develop in later years. Judging by the number of abandoned radiation units (at least fifty in Brazil alone), there will certainly be more such tragedies in the future.

The military also has nuclear accidents, but most information about them is secret. In the United States accidents are classified, in order of increasing severity, as Dull Swords, Bent Spears, Broken Arrows, or, worst of all, Nucflash – a nuclear accident that could lead to war. The Department of Defense has acknowledged at least 27 Broken Arrows between 1950 and 1981. All involved the Air Force. On at least two occasions when planes carrying nuclear bombs crashed, the bombs were destroyed (though they did not explode) and plutonium was scattered over a wide area. The Navy has also acknowledged that it has several hundred pages of documents under the heading 'Summary of Navy Nuclear Weapons Accidents and Incidents', covering the period from 1973 to 1978. The Army claims never to have had a serious accident with nuclear weapons. The Stockholm International Peace Research Institute says that there were 125 accidents involving US nuclear weapons between 1945 and 1976 – just over one every three months. Military nuclear accidents in other countries are even harder to document than American ones, but Western intelligence services have details of several Soviet accidents, including, in the late 1960s, 'a reactor casualty' aboard the *Lenin*, a nuclear-powered icebreaker. The CIA has admitted the existence of a still-classified paper, called 'Submarine Accidents: A Continuing Problem for the Soviet Navy'.

Nuclear accidents in space sounds like the title for a grade B horror movie, but there have already been several accidents involving nuclear power units in space satellites, both Russian and American. Plutonium–238 has been used to supply power for scientific experiments and communications equipment on board American space satellites since 1961. These units are called space nuclear auxiliary power (or SNAP) devices. On 21 April 1964, a satellite carrying a unit known as SNAP 9–A was unsuccessfully launched. As it re-entered the atmosphere over the Indian Ocean, the satellite and its SNAP unit disintegrated. SNAP 9–A contained 17,000 curies of plutonium, which returned to earth within several years, and is deposited all over the globe. The plutonium released by this one accident equals five per cent of the plutonium released by all atmospheric weapons tests combined.

Four years later, when the Nimbus-BI spacecraft was deliberately destroyed due to a guidance error, its SNAP device fell into the Santa Barbara channel off the coast of California and was recovered intact. In 1970 when the Apollo 13 spacecraft was damaged, a SNAP device, attached to the lunar module, fell into the South Pacific, near the island of Tonga. The device has never been recovered, but officials say that radiation levels in the area indicate that no plutonium was released.

In 1978, a nuclear-powered Russian satellite, Cosmos 954, disintegrated as it re-entered the atmosphere over Canada. A 600-mile-long strip of northwest Canada was strewn with radioactive debris. Despite an exhaustive search by the Canadian government, it is estimated that less than one per cent of the radioactive deposits were found. The total amount of radioactivity released is unknown, partly because the Soviet Union has refused to divulge any technical information about the incident.

Another Soviet satellite, Cosmos 1402, launched in 1982, carried a nuclear reactor that was intended to be boosted to a higher orbit when its job was done. Instead, in 1983, the reactor re-entered earth's atmosphere. Scientists believe that the reactor was destroyed by the atmospheric friction and is gradually returning to earth as fine radioactive particles. High-altitude balloons launched by the US Department of Energy and equipped with paper filters have detected 'a clearly measurable excess of uranium–235' from the Russian space reactor in the atmosphere. These five incidents occurred out of a total of just over 40 space flights involving nuclear devices, for an accident rate of more than one in ten.

The next US space mission involving plutonium is Project Galileo, a space probe designed to explore the planet Jupiter. Galileo, delayed by the fatal Challenger crash in January 1986, and now scheduled for 1989, will carry 280,000 curies of plutonium–238 (almost 50 pounds of plutonium). According to Merril Eisenbud, 'this is a huge quantity . . . which requires maximum assurance that containment of the fuel will be achieved in the event of any of the many types of accidents that can be hypothesized'. A 1987 NASA analysis found that the plutonium is unlikely to be released in a launch-pad accident because it will be shielded in material which can 'survive pressures in excess of 2,000 psi (pounds per square inch)'. An earlier Department of Energy analysis had concluded that certain types of accidents could generate pressures of up to 19,600 pounds per square inch, but NASA asserts that this conclusion is incorrect. Once the shuttle is launched, the Galileo probe will take a complicated route to Jupiter: it will return towards Earth and circle the globe twice, using Earth's gravitational field to build up its momentum to carry it towards Jupiter. The 1987 NASA report found that 'there is a remote but finite chance that the spacecraft may re-enter the Earth's atmosphere . . . less than one chance in one million'. The report also promises that 'procedures will be developed to minimize the possibility of inadvertent Earth re-entry'.

Epilogue

Lauriston Taylor has described the history of radiation protection as a pattern of lulls and spurts. The lulls, he wrote, are caused by 'ignorance and indifference' rather than negligence. 'The spurts have almost invariably been brought about by painful experience and the sudden but too tardy realization that [the use of ionizing radiation] has offered to humanity another sacrifice, another martyr to medical science.'

With the evidence from the Hiroshima and Nagasaki bombings indicating that the risk estimates on which the ICRP has based its exposure limits are from five to ten times too low, it is time for another spurt. Indeed, at its 1987 annual meeting, the ICRP acknowledged that the risk from radiation exposure is greater than previously realized. It appears certain that the commission will lower its exposure limits for radiation workers and possibly also for members of the public. The question is, what will the new limits be, and when will they take effect?

While acknowledging that its dose limits are based on erroneous risk estimates, the ICRP has decided not to lower those limits yet. It offers several reasons for this. One is that it wishes to wait to hear from UNSCEAR, the United Nations scientific committee, and the US Academy of Sciences, both of which are presently revising their risk estimates. The commission also argues that there is no hurry to lower dose limits since 'the requirement to keep all doses "as low as reasonably achievable" [the ALARA concept] should in most situations keep doses far below the dose limits'.

Only after UNSCEAR has issued its revised risk estimates will the ICRP issue its new recommendations. These, says the commission, 'are expected to be completed by 1990'. How much longer it will be before national governments adopt them is anybody's guess. The ICRP's current recommendations, issued in 1977, were so complicated that it was many years before they were put into practice. Britain took eight years to incorporate them into its legal system. Many countries, including the United States, are still using the 1956 standards. One reason for the delay in adopting the 1977 recommendations is that the ALARA concept, adopted as a substitute for lowering the dose limit, is ambiguous and difficult to translate into law. If the ICRP

now decides to simply reduce the existing dose limit, its recommendations would probably be promptly adopted by most countries. The
National Radiological Protection Board, the British government's
adviser on radiation safety, has already acted. In 1987, after the ICRP
announced that its dose limits would remain unchanged until at least
1990, the NRPB decided to take matters into its own hands. It directed
that the dose limit for workers be immediately reduced from 5 rems per
year to 1.5 rems per year. This is the first time that a governmental
radiation protection body has acted in advance of the ICRP.

That the commission will have to cut its existing dose limits seems
certain. How large those cuts will be, no one knows. At present, the
commission estimates that its risk estimates are off by 'a total factor of
the order of two'. Individual members of the ICRP have gone further,
however. John Dunster has predicted 'a decrease in dose limits by a
factor perhaps as high as five'. And another British member of the
commission, Roger Berry, has said that ICRP's risk estimates are from
two to five times too low. The NRPB's new occupational dose limit of
1.5 rems per year reflects a feeling that existing risk estimates are
about three times too low. The evidence from the Radiation Effects
Research Foundation indicates that the ICRP may be underestimating
risk by a factor of ten or more.

For the first time in its history, the ICRP is under pressure to admit
the views of outsiders. Before the commission's 1987 meeting, Friends
of the Earth presented commission chairman Daniel Beninson with a
petition, signed by 800 scientists from 16 countries, in which was set
forward a number of arguments for changes in the commission's
recommendations. Among the petitioners' points were that dose limits
for workers and the public should be lowered immediately by a factor of
five; and that safeguards should be introduced to guard against the
accidental irradiation of women in the early months of pregnancy when
the foetus is particularly vulnerable to mental retardation. The most
radical demand, however, was for 'worker and public representation on
the commission'.

At first glance, it would seem absurd for a scientific body dealing
with highly technical matters to include laymen. But the ICRP is not
strictly a scientific body. It does not simply produce risk estimates or
study the effects of radiation on humans and the environment. The
ICRP, in fact, largely relies on other sources for such information. Its
unique contribution is to take the risk estimates produced by others and
to tell governments and industries how much risk their citizens and

workers should be subject to. This is a value judgement that the citizens and workers are eminently qualified to make. In fact, such representation would be much in the spirit of the original ICRP which began as a group of doctors setting protection standards for themselves and their fellow doctors. The concept is a democratic one: no radiation without representation. Radiation is a scientific matter, but radiation protection is more than that. It is, as Lauriston Taylor has eloquently pointed out, a 'problem of philosophy, morality, and the utmost wisdom', and these are qualities on which science does not claim to have a monopoly.

Chronology of Radiation Exposure Limits

Workers

1934 0.1 roentgen per day (approximately 30 rem per year) – ICRP
1950 0.3 rem per week (15 rem per year) – ICRP
1956 5 rem per year – ICRP
1977 no more than 5 rem per year, but 'as low as reasonably achievable', (the ALARA principle) – ICRP
1987 1.5 rem per year – NRPB (UK)

Radon Standards

1941 10 picocuries per litre of air – Manhattan Project
1953 100 picocuries per litre of air (12 Working Level Months per year) – Tripartite Conference
1959 30 picocuries per litre of air (3.6 Working Level Months per year) – ICRP
1971 33 picocuries per litre of air (4 Working Level Months per year) – EPA (US)
1981 40 picocuries per litre of air (4.8 Working Level Months per year) – ICRP

Individual Members of the Public

1949 0.3 rem per year (one per cent of occupational limit) – Tripartite Conference
1953 1.5 rem per year (10 per cent of occupational limit) – Tripartite Conference
1954 1.5 rem per year (10 per cent of occupational limit) – ICRP
1956 0.5 rem per year (10 per cent of occupational limit) – ICRP
1985 0.1 rem per year, but exceptions up to 0.5 rem per year – ICRP

Population Limits

1959 0.17 rem per year – ICRP (made obsolete by stricter standards for individual members of the public, see above)

Glossary

Types of Radiation

Alpha particle Positively-charged ionizing particle emitted by some radioactive substances.

Beta particle Negatively-charged ionizing particle emitted by some radioactive substances.

Electromagnetic radiation Waves or particles of energy which vary according to the amount of energy each type emits. At the high-energy end of the electromagnetic spectrum are the ionizing radiations, including X-rays, gamma rays and alpha, beta and neutron particles. Lower-energy, non-ionizing radiations include micro waves, radio waves and light waves.

External radiation Radiation from a source outside the body. The term is usually used about radiations that can penetrate human skin and thus can cause biological damage from outside the body such as gamma and X-rays.

Gamma ray Highly penetrating ionizing radiation emitted by some radioactive substances.

Internal radiation/Internal emitters Radiation from a source inside the body. These terms are usually used in reference to radioactive substances that emit alpha or beta particles which, because they cannot penetrate human skin, are most hazardous once they are inside the body.

Ion Atomic particle or atom bearing an electric charge, either positive or negative.

Ionization The process by which a neutral atom acquires a positive or negative charge.

Ionizing radiation Radiations at the high-energy end of the electromagnetic radiation spectrum. When they pass through matter, ionizing radiations are capable of disrupting the structure of atoms by separating electrons from the rest of the atom (ionization).

Isotope Isotopes are atoms that are identical except that they have different numbers of neutrons. Nearly all elements in nature consist of several different isotopes.

Neutron Uncharged ionizing particle emitted by cosmic radiation and nuclear reactors.

Neutron activation products Substances that are made radioactive by being bombarded by neutrons.

Nuclear radiation Radiation emitted from the nuclei of atoms during radioactive decay, nuclear fission, or bombardment with electrons.

Radioactive substance A substance that emits ionizing radiation through a natural process known as radioactive decay.

Radioisotopes Radioactive isotopes, or radioisotopes, are isotopes of an element that is radioactive.

Radionuclides Any atom that is radioactive.

Radon daughters Decay products of radon.

Roentgen ray An early name for X-ray.

X-ray Highly penetrating radiation, similar to gamma rays, but produced by machines, not emitted by radioactive substances.

Radiation Measurement Units

Collective dose The sum of all the radiation doses received by all exposed persons in a given population.

Curie (Ci) A measurement of radioactivity. One curie undergoes 37 billion atomic disintegrations per second. This is the amount of radioactivity present in one gram of radium-226.

Erythema dose An early and inexact measurement of radiation, no longer in use. A erythema dose is the amount of X or gamma radiation necessary to create a visible reddening of an exposed person's skin.

Person-rem A measurement of collective dose, the total number of rems received by a population.

Rad A measurement of the amount of radiation received by a person. Rad stands for Radiation Absorbed Dose.

Rem A measurement of the biological effect of radiation. This unit takes account of the fact that different forms of ionizing radiation have different biological impacts. A given amount of alpha radiation, for example, is approximately ten times more biologically effective than the same amount of gamma radiation. Thus, one rad of alpha radiation translates into ten rems, whereas one rad of gamma radiation equals only one rem.

Roentgen (r) A unit that measures radiation by its power to ionize a given amount of air, now seldom used. For X and gamma rays, one r is equivalent to one rem.

Working level (WL) A measure of radiation exposure to alpha radiation in air, usually used to measure radon gas exposure in mines.

Working level month (WLM) The radiation dose resulting from inhaling air contaminated by one working level of radiation for one working month (170 hours).

Note: A new system of measuring radiation has been adopted, but it is not yet in universal use. The old system is still used by the United States government, among others. For consistency's sake, I have used the old system throughout the book. Here is a table of equivalencies for those who wish to translate.
1 curie = 37,000,000,000 Bequerels.
1 rem = 0.01 Sievert
1 rad = 0.01 gray.

Other Terms

Cathodograph An early term for an X-ray picture.

Critical mass The amount of radioactive material required to sustain a nuclear reaction.

Fluoroscope An X-ray machine that projects images onto a screen instead of taking still photographs.

Genetic injury An injury to a person's genes, which can be passed on to subsequent generations.

Half life The length of time it takes for half of a given quantity of a radioactive substance to decay to stability.

Nuclear fission The splitting of an atom's nucleus into several parts, accompanied by the release of massive amounts of nuclear energy.

Somatic injury An injury to the body of the exposed person which is not passed on to succeeding generations.

Metric System

Prefix	Symbol	Value
mega	M	1,000,000
kilo	k	1,000
milli	m	1/1,000
micro		1/1,000,000
nano	n	1/1,000,000,000
pico	p	1/1,000,000,000,000

Notes

1 Discovery

5 The story of Roentgen's discovery of X-rays is told by Glasser, Bleich, Brecher, and Donizetti.

6 Edison's involvement with X-rays is detailed by Josephson and Brecher.

6 farmer in Iowa: Mould, p. 34.

6 drunkards and smokers: Kathren (1979), p. 34.

6 criminal behaviour: Kathren (1984), p. 12.

6 'learning anatomical details': Bleich, p. 6.

6 dog salivate: Kathren (1979), p. 34.

6 Thurston Holland's lecture: Mould, p. 44.

7 'naughty, naughty Roentgen Rays.': Bleich, p. 6.

7 banning the use of X-rays in opera glasses: Ibid.

7 X-ray proof underwear: Mould, p. 34.

7 'fascinating and coquettish' X-ray: Hilgartner et al, p. 2.

7 doctors . . . rushed to experiment: Brecher, p. 25.

7 Within a year: Badash (1965), p. 134.

7 John Cox: Brecher, pp. 18–19.

8 Fuchs: Ibid., p. 64.

8 Freund: Mould, p. 4.

9 Clarence Dally: Brown, pp. 32–42.

9 Herbert Hawks: Brecher, pp. 82–83; Brown, pp. 9–10.

10 Patients were burned: Dewing, p. 110.

11 'even go out for lunch': Taylor (1971), p. 82.

11 23 severe X-ray injuries: N. S. Scott, 'X-ray Injuries', *American X-Ray Journal*, 1, (1897) pp. 57–65.

11 Elihu Thompson: Brown, pp. 11–12.

12 One injury in 1901: Brecher, p. 160.

2 The First Standards

13 Clarence Dally: Brown, pp. 32–42.

13 'agony of inflamed X-ray lesions': Brecher, p. 165.

13 allegorical figure of Science: Kathren (1962), p. 507.

14 Rollins' seven-minute exposure guideline: Brecher, p. 182.

14 Charles Leonard: Ibid., p. 178.

14 'no such thing as absolute protection': Ibid., p. 179.

14 Caldwell: Ibid., pp. 179–180.

14 'attempts to ignore . . . effects of X-light on animals': Rollins, quoted in Hilgartner et al., p. 9.

15 Kassabian: Brown, p. 92.

15 Brown's death: Brecher, p. 170.

15 'undesirable risks': Ibid., p. 175.

15 '100 named diseases': Ibid., p. 146.
15 radiation treatments for women: Hilgartner et al, pp. 6–7.
16 Tricho treatments: Lapp and Schubert, pp. 101–103.
17 ICR recommendations, 1928: Taylor (1979), p. 3–022.
18 erythema dose could vary by 1,000 per cent: Taylor, op. cit., p. 3–013.
18 'wait until the skin began to redden': Taylor, interview with author, 8 August, 1984.
18 different radiation units: Mould, p. 11; Taylor (1979) p. 3–010.
19 Mutscheller's research: Mutscheller, pp. 65–70.
19 no more than six hospitals: Taylor (1984), interview with author.
19 'not yet sufficiently checked': Mutscheller, p. 67.
19 Rolf Sievert's research: Taylor (1979), p. 3–012.
20 Barclay and Cox: Ibid.
20 'Pure judgement': Taylor, 'Judgement in Achieving Protection Against Radiation', IAEA vol. 22, No 1, p. 17.
20 Meyer and Glasser: Taylor (1979), p. 3–013.
20 Kustner: Taylor (1979), p. 3–015.
20 common to round the ED up to 600 R: Taylor (1971), p. 86.
21 US committee adopted 0.1 r per day: Taylor (1971), p. 87.
21 'an unreasonable knowledge': Taylor (1979), p. 4–015.
21 International committee adopted 0.2 r per day: Taylor (1979), p. 87.
21 'We just didn't see any difference': Taylor (1984), interview with author.
21 'Mutscheller's work was seriously flawed': Ibid.
21 Monument to the 'martyrs of radiation': Mould, p. 5.

3 Radioactivity

22 Becquerel's discovery is described in Kathren (1979), pp. 21–22; in the *Encyclopedia Britannica*, 15th edition (1982), p. 434; and in Christine Sutton, 'Serendipity or sound science', *New Scientist*, 27 February, 1986, pp. 30–32.
23 The story of Marie Curie's discovery of radium is told in Brodsky, ed., (1979) pp. 22–23; in Eva Curie's biography of her mother; and in Peter Craig, 'The light and brilliancy of Marie Curie', *New Scientist* (26 July, 1984), pp. 32–35.
25 lighting bicycles with disks of radium: Badash, p. 21.
25 mixing radium with chicken-feed: Badash, p. 27.
25 radium-fertilized soil: Hilgartner et al, p. 6.
25 Loie Fuller and the Curies: Curie, pp. 232–233.
25 Becquerel's radium burn: Brecher, pp. 155–156.
26 Alexander Graham Bell: quoted in James T. Case, 'Early History of Radium Therapy and the American Radium Society', *American Journal of Roentgenology*, 82 (1959), p. 578.
26 $120,000 per gram: Quimby, p. 444.
26 'Radium has absolutely no toxic effects': Quoted in Lapp and Schubert, p. 112.
26 'At first we gave very large doses': Brecher, p. 291.
26 uses of radium: Evans (1974), p. 499; Lapp and Schubert, p. 113.
26 Dr Hall-Edwards: Mould, p. 6; Hilgartner et al, p. 8.
27 radium treatments at Memorial Hospital: Brecher, p. 277.
27 number of hospitals and doctors with supplies of radium: Brecher, p. 279.
27 fewer than 50 physicians: del Regato, Juan, A., 'The American Society of Therapeutic Radiologists', *Cancer*, Vol. 29, No. 6 (June 1972), p. 1443.
27 'The disastrous effects of over-exposure': Brecher, p. 279.
28 descriptions of various 'radium products' are contained in *New Scientist* (25 September, 1986), p. 88; Lapp and Schubert, p. 112; Kathren (1979), pp. 35–6; Kathren (1984),

pp. 13–14; Dewing, p. 115; Michel Mok, 'Radium', *Popular Science* Vol. 121, No. 1 (1932), p. 11.

28 Byers: Evans (1984), p. 579; Kathren (1984), p. 13; Lapp and Schubert, p. 112; Mok, op. cit., p. 9.

28 an article in *Life* magazine: *Life* (July 7, 1952).

28 'nature's own remedy': publicity brochure for Merry Widow Mine, 1985.

28 'more visits to the mine': Ibid.

4 The Dial Painters

29 'weirdly artistic effect.': Sochocky, S. A., 'Can't You Find the Keyhole?', *The American Magazine* (January 1921), p. 24.

30 One in six American soldiers: Ibid., p. 26.

30 'wiping the brush clean between their lips': Evans, R., 'Radium Poisoning: A Review of Our Present Knowledge', *American Journal of Public Health*, 23 (1933), p. 1019.

30 This account of the radium dial painters and those who helped solve the mystery of their deaths is taken from a number of sources, including: Cloutier, Roger, 'Florence Kelley and the Radium Dial Painters', *Health Physics Journal* Vol. 39, No. 6 (1980), pp. 711–717.
Evans (1974).
Martland, Harrison S., 'Occupational Poisoning in Manufacture of Luminous Watch Dials', *Journal of American Medical Association* (1929) Vol. 92, No. 6 pp. 466–473.
Lang (1959).
Goldmark, Josephine Clara, *Impatient Crusader*, University of Illinois Press, Urbana (1953), pp. 189–204.

33 'intravenous injections . . . of radioactive treatments sanctioned by the American Medical Association': Lapp and Schubert, pp. 113 and 116; Evans (1974), p. 499.

5 Setting Limits on Radium

38 'too great reliance is not to be put on the above figures': Taylor (1979), p. 4–007.

39 'Several hundred times more resistive': Taylor, op. cit., p. 5–024.

39 'There was no pressure': Evans, telephone interview with author, 14 Jan, 1986.

40 The setting of the first standard for internal exposure to radium is described in Taylor (1979), pp. 5-022–5-025, and Evans (1984), pp. 579–581.

40 'If it hadn't been for those dial painters': Lang, p. 51.

6 Building the Bomb

43 'It suddenly occurred to me': Weart and Szilard, p. 17.

43 Szilard's patents: Ibid., p. 18.

45 'my assumption . . . was really absurd': Jungk, pp. 61-62.

45 'horrifying conclusion': McKay, p. 26.

45 '15 March, 1939': Ibid., p. 34.

45 more than 100 scientific papers: Ibid., p. 35.

45 'War itself will become obsolete': Langer, R. M., 'Fast new World', *Colliers*, July 6, 1940, p. 18.

46 The Maud Ray incident is described by Margaret Gowing in *Niels Bohr, A Centenary Volume*, edited by A. P. French and P. J. Kennedy (Harvard University Press, 1985), p. 266.

47 The genesis and operation of the Manhattan Project are described in Compton, Clark, Groueff, Jungk, Hacker, Lamont, McKay, and Stone.

47 'they say it is a very pleasant way to die': Wyden, Peter, *Day One* (New York: Simon and Schuster, 1984), p. 345.

47 'our physicists became worried.': Compton, p. 177.

48 proposal to restrict radiation exposure: Taylor (1979), pp. 5–013–21.

48 'just as arbitrary': Ibid., p. 5–016.

48 'lost in another crack': Taylor, interview with author, 8 August, 1984.

48 'rather poor experimental evidence': Stone (1946), p. 17.

48 'even this factor is too large': Stone (1946), p. 17.

49 'definite biological changes': Hacker (m.s.) II p. 38

49 'untoward psychological effects': Ibid., II p. 38

49 monitoring equipment was scarce and rather primitive': Stone (1946), p. 14.

49 'We do not have important legal evidence': Hacker (m.s.) III p. 17

50 'dangerous changes months or years after': Hacker (1987), p. 55.

50 'one vast experiment': Hacker (m.s) II p. 21.

50 strengthen 'the Government interests': Hacker (1987), p. 51.

50 Stone wanted long-term research: Ibid., pp. 50–52.

50 people suffering from incurable diseases: Hacker (m.s.) II pp. 23–24.

51 'nature and extent of the hazards . . . were not known': Stone (1951), pp. 1–2.

52 'none at all': Hacker (m.s.) II p. 44.

53 revised plutonium standard: Hacker (1987), pp. 62–3.

53 'immediate high amputation': Bradley, p. 134.

53 secret insurance fund for plutonium chemists: Hacker (1987), p. 62.

54 'more radiation dosage than was desirable': Hacker (m.s.) III p. 20.

54 'two major plutonium spills': Hacker (1987), p. 66.

54 'ten milligrams of plutonium exploded': Ibid., p. 67.

54 'some workers already had too much': Ibid., p. 68.

54 'impossible to keep all sources of radiation under control': Stone, p. 14.

54 'an entire building uninhabitable': Hacker (m.s.) II pp. 44–5.

54 'the employees got down on their knees': Lang (1948), p. 51.

55 'Contamination was rampant': National Committee for Radiation Victims, p. 19.

55 Ted Lombard and the VA: Bertell, pp. 65–69.

56 'there were no over-exposures': Hacker (m.s.) II p. 45.

56 'no one lost his life': Hacker (1987), p. 57.

56 'a larger medical organization': Hacker (m.s.) II p. 42.

56 'a brief indoctrination': Ibid., Hacker (m.s.) II p. 42.

7 Getting to Know the Bomb

57 'The idea was to explode the damned thing': Hacker (1987), pp. 84–5.

57 'new calculations': Ibid., p. 90.

58 'the individual's responsibility' to limit his own exposure': Ibid., p. 89.

58 'half of us': Ibid., p. 91.

58 'even 100 r would not be harmful': Ibid., p. 93.

59 'You boys must have been up to something': Ibid., p. 102.

60 'They were just roasting the meat': Ibid., p. 103.

60 the ground appeared 'frosted': Ibid., p. 105.

61 fallout in the Wabash River: *Time* (November 12, 1945), p. 50. Eisenbud, p. 274; Lang, D., *From Hiroshima to the Moon*, p. 367.

62 'the basic power of the universe': *Time* (August 13, 1945).

62 'Hiroshima will be a devastated area': Boyer, p. 188.

62 'our mission was to prove that there was no radioactivity': Hacker (1987), p. 110.

62 'below the hazardous limit': Ibid., p. 112.
63 'I Write This As A Warning To The World': Burchett, Wilfred, *Shadows of Hiroshima* (London: Verso Editions, 1983), pp. 8 and 34.
63 'No Radioactivity in Hiroshima Ruin': *New York Times* (September 13, 1945).
63 'radiation was not mentioned': Boyer, p. 307.
63 'just another piece of artillery': Clarfield and Wiecek, p. 81.

8 Postwar Standards

64 'The atom had us bewitched': David Lilienthal, *Change, Hope, and the Bomb* (Princeton, N.J.: Princeton University Press, 1963), p. 18.
64 'We could face doomsday': *Woman's Home Companion* (May 1948), p. 32.
64 'Beyond the veil of dust and smoke': *The New York Times* (August 9, 1945), p. 20.
64 'the possible benefits . . . must be emphasized': Boyer, p. 124.
64 'Buck Rogers': *Newsweek* (August 20, 1945), p. 59.
65 'force him back into the cave': Mazuzan and Walker, p. 2.
65 'Among the ideas mooted': Ibid., p. 2; Boyer, pp. 112–113.
65 'it is unlikely that you or your classmates will die prematurely': Boyer, pp.110–111; 119–120.
65 'setting off a series of bombs in the Arctic': Ibid., p. 111.
65 'No baseball game will be called off': Dietz, David, *Atomic Energy in the Coming Era* (New York: Dodd, Mead, and Co., 1947), pp. 16–20.
65 'it was time to revive the U.S. Advisory Committee': Taylor, Lauriston, (1979), p. 7–001. Much of the following account of the NCRP's postwar years is taken from Taylor's official history.
66 idea of including insurance companies 'did not find any enthusiastic response': Ibid., p. 7–004.
66 asked the committee to prepare a handbook for embalmers: Ibid., p. 7–007.
66 so that companies could offer atomic insurance: Ibid., p. 7–057.
67 what dose of atomic bomb radiation would be lethal: Ibid., p. 7–013 to p. 7–015.
67 'These considerations affect collective bargaining': Ibid., p. 7–059.
67 'I see no alternative but to assume that the operation is safe': Ibid., p. 7–061 to p. 7–062.
67 the AEC 'was putting considerable pressure on': Ibid., p. 7–010.
68 'demonstrated that we could do something for them': Ibid., p. 7–013.
69 'so we lowered it': Lauriston Taylor, interview with author, August 8, 1984.
69 'it is just this group . . . that we are most interested in protecting': Taylor (1979), p. 7–181.
69 'a rather high incidence of abnormalities': Ibid.
70 'give him the final say': Ibid., p. 7–182.
70 'part of the many other risks that accrue to military personnel': Ibid., p. 7–124.
71 A few per cent may be five times more sensitive to radiation: Eisenbud, p. 42.
71 'From the public relations angle': Taylor (1979), p. 7–051.
71 'This had certain advantages': Ibid., p. 7–055.
71 'never able to find a cut-off point': Ibid., p. 7–119.
71 'We cannot be forever postponing action': Ibid., p. 7–013.
72 'a ring of authority': Ibid., p. 7–182.
72 'emphasize that these were interim values': Ibid., p. 7–153.
72 nuclear plants should be able to meet more stringent standards in future: Ibid., p. 7–152 to p. 7–153.
72 'it might be some years before our knowledge . . . was sufficiently clarified': Ibid., p. 7–127.
73 'The only qualified protection experts . . . are already members': Ibid., 737–080.
73 'might put some of us in an embarrassing situation': Ibid., p. 7–090.
73 response did not mollify Parker: Ibid., p. 7–092.

73 NCRP accepted 'no-threshold' theory: Taylor (1980), p. 18.
74 panic in the event of an atomic explosion: Taylor (1979), p. 7–124.
74 risks were outweighed by the benefits: Ibid., p. 7–123.

9 Uranium Fever

75 'every butterfly catcher': Dean, Gordon. *Report on the Atom* (New York: Knopf, 1963), p. 38.
75 'a worthless metal': Taylor and Taylor, p. 79.
75 a $10,000 discovery bonus: Atomic Energy Commission (1951).
76 Paddy Martinez's strike: Lang, Daniel. *From Hiroshima to the Moon*, p. 305.
76 'Get Rich From That Miracle Atom': Taylor and Taylor, p. 3.
76 'Among the uraniumaires': Kursh, Harry, *How To Prospect for Uranium* (Greenwich, Connecticut: Fawcett Books, 1955), p. 6.
76 30 million uranium mine shares were sold: Taylor and Taylor, p. 25.
76 'No Talk Under $1,000,000': Ibid., p. 5.
77 the Mickey Mouse Club: Ibid., p. 116.
77 hundreds of uranium mines: Moss, p. 30.
77 one-quarter of the country's known uranium reserves: Taylor and Taylor, p. 309.
77 the average uranium miner gets a dose of 200 millrems per year: Brooks, Table 5.
78 'the interior of the mine was a nasty area': US Senate Special Committee on Aging, p. 77.
78 'The average age of the miners who die'. Rashke, Richard *The Killing of Karen Silkwood* (Boston: Houghton Mifflin, 1981), p. 46.
78 According to the US Public Health Service: Health Research Group et al (1980), p. 6; and Wagoner (1980), p. 4.
78 'This thing comes back every day': Phillip Harrison, interview with author, April 26, 1985.
79 the 72-acre tailings pile: Department of Energy, *Fact Sheet: Shiprock Uranium Mill Tailings Site* (August 20, 1982), p. 1.
79 'National Sacrifice Areas': National Academy of Sciences. *Rehabilitation Potential of Western Coal Lands* (Cambridge, Mass: Ballinger Publishing Co., 1974), pp. 85–86.
80 'We had no idea at all': Phillip Harrison, interview with author, April 26, 1985.
80 compensation laws are not designed to deal with . . . radiation-induced damage: Stanton, p. 9, and *Navajo Times* (September 7, 1978), pp. 8–10.
80 'I have seen a lot of buckpassing': US Senate Special Committee on Aging (1980), p. 16.
80 'the most well-documented case of an occupational disease', Health Impact of Low-Level Radiation (1979), p. 23.
81 The story of the Erzgebirge miners is told in Morgan, J. R., 'A History of Pitchblende', *Atom: the Journal of the United Kingdom Atomic Energy Authority* (March, 1984), pp. 63–68.
81 Studies of 'miner's disease' are described in Health Research Group et al (1980), p. 5.
81 cancer among plitchblende miners was being cited as the classic example: Wagoner (1981), p. 3.
81 the French were installing ventilation systems: Wasserman et al, p. 148.
82 'the radioactivity . . . is not dangerous to humans': AEC and USGS (1951).
82 only two laboratories in the entire country: Health Impact of Low-Level Radiation (1979), pp. 21 and 26.
82 deemed not to apply to mines: US Congress Joint Committee on Atomic Energy (1967), p. 278.
82 'we were not permitted to do so': Health Impact of Low-Level Radiation (1979), p. 21.
82 the more lenient limit . . . became the US standard: Kathren, p. 126.
83 In 1952 . . . the average uranium mine: Puchin, p. 153.
83 exposed to more than 10 WLs: Wagoner (1981), p. 5.

83 'was effective in reducing average ... values': *US Uranium Registry*, 'Occupational Exposure to Uranium: Processes, Hazards, and Regulations' (1981), p. 30.

83 'as great or greater than the European experience': US Senate Special Committee on Aging (1980), p. 58.

83 'It was in 1962 that we first showed ... an increased risk': Wagoner (1980), pp. 4–5.

84 'the primary objective of the FRC': JCAE (1967), p. 34.

84 'It is hard to believe that': Health Impact of Low-Level Radiation (1979), p. 23.

84 'six hundred to eleven hundred will die': Hertsgaard, p. 155.

84 'From twelve to four': John Parker, interview with author, April 23, 1985.

85 more than 3,000 WLMS: Health Research Group, p. 5.

85 'well below the internationally accepted limits': Rippon, p. 5.

85 well under the limit: Wagoner (1980), p. 3.

85 a monitoring 'blitz': Wagoner (1981), p. 5.

86 'to exceed federal exposure standards': Mining Safety and Health Administration, *Mine Regulation and Productivity Report*, Vol. 2, No. 43 (November 17, 1978), p. 2.

86 205 of the 3,400 ... had died: Health Research Group, p. 6.

86 30,000 to 40,000 men have worked in underground uranium mines: US Senate Special Committee on Aging, p. 85.

86 'Nobody never did say anything': James Aloyisius, interview with author, April 26, 1985.

86 'Did you ever like to sneak a cigarette?': Red Souther, interview with author, April 23, 1985.

87 smoking does increase a miner's risk: Health Research Group, p. 5 and Wagoner (1980), pp. 7–9.

87 companies continue to tell them that ... they will not develop lung cancer: Wagoner (1980), p. 7.

87 'this epidemic': Health Impact of Low-Level Radiation (1979), p. 31.

87 'you might as well guess': US Congress Joint Committee on Atomic Energy (1967), p. 283.

87 'from 0.1 WL to 0.3': Ibid., p. 282.

87 officials believed, wrongly: Ibid.

88 would not have a significantly greater risk: Kathren, p. 127.

88 studies ... have shown excess fatal cancers: Wagoner (1980), p. 8; Health Research Group, p. 8; and Public Citizen Litigation Group (1985), pp. 16–22.

88 'cannot be characterized as safe': Public Citizen Litigation Group (1985), p. 21.

88 filed a petition: Health Research Group and Oil, Chemical, and Atomic Workers International Union, 'Petition Requesting An Emergency Temporary Mandatory Standard For Radon Daughter Exposure' (1980).

88 promised a 'serious review': Public Citizen Litigation Group (1985), p. 11.

88 'was not sufficiently protective': Larry Mazzuckelli, telephone interview with author, June 6, 1985.

10 Operation Crossroads

89 Accounts of Operation Crossroads can be found in Bradley, Hacker, Warren (1947), Makhijani and Albright, and *Operation Crossroads: The Official Pictorial Record*.

89 'The islanders ... are well pleased': Rafferty et al, p. 29.

90 'their psycho-neurotic tendencies': *Operation Crossroads*, p. 67.

90 'their votes were needed in Washington': Hacker (1987), p. 124.

90 'to reassure Col. Warren that ... no successful suits could be brought': Makhijani, p. 20.

90 'literally a hundred times larger': Ibid., p. 21.

91 military aims would take precedence: Hacker (1987), p. 117.

91 'The Good That May Come From The Tests' *Newsweek* (July 1, 1946), p. 21.

91 the humanitarian aspect: *Science News Letter* (May 11, 1946), p. 294.

91 'atomic playboy': Rafferty et al, p. 27.
92 'highly overrated' Bradley, p. 58.
92 'Awful as it was': *Time* (July 8, 1946), p. 20.
92 'finite after all': *Time* (July 15, 1946), p. 31.
92 'the average citizen is now only too glad to grasp as the flimsiest means', *New York Times* (August 4, 1946), p. 3.
92 'tons of radium': Hacker (1987), p. 125.
93 'no . . . physiologically significant radiation dose': Ibid., p. 135.
93 'a perfect test': Ibid., p. 136.
93 'A gigantic flash – then it was gone': Bradley, p. 92.
93 'We were thankful for that': Ibid., pp. 96–7.
94 'Radioactivity persisting in water and on board': Hacker (1987), p. 139.
94 'boarding inspection may be dangerous for weeks': Makhijani, p. 10.
95 'costume in the tropics': Bradley, p. 131.
95 'order a whole bunch of sailors off a boat': Bradley, interview with author, July 26, 1986.
95 'hairy-chested' approach to radiation: Hacker (1987), p. 144.
95 'an attitude of indifference': Ibid.
95 'The target vessels are . . . extensively contaminated': Makhijani, p. 21.
96 'I'm getting the jitters': Hacker (1987), p. 147.
96 'many will certainly be sunk': Bradley, p. 115.
96 'the water was steadily increasing in radioactivity'. Ibid., p. 101.
97 'a small but definite amount of plutonium': Ibid., p. 147.
97 'unanimous refusal': Makhijani, p. 20.
97 Bradley was able to compute the dose he received: Bradley, interview with author, July 26, 1986.
98 'increasing amounts day by day': Makhijani, p. 10.
98 'the evaporators . . . have become quite hot': Bradley, p. 103.
98 'I drank milk instead of water': Bradley, interview with author, July 26, 1986.
99 'they might as well continue to operate': Makhijani, p. 13.
99 'To the sailors, it must all be incomprehensible': Bradley, p. 145.
99 worried that they might have eaten contaminated food: Saffer, p. 209.
99 'two were spent vomiting and retching': Wasserman et al, pp. 43–44.
99 'He didn't advise us any decontamination procedure': Ibid., p. 44.
100 'I discovered that I hadn't escaped': John Grifalconi, interview with author, July 24, 1986.
100 'very much afraid of claims being instituted': Makhijani, p. 20.
100 'no casualties from overexposure to radiation': Hacker (1987), p. 153.
100 'the spiritual effect was great': Jungk, p. 246.

11 The Proving Ground

101 'will remain contaminated for years': Warren, Stafford, 'Conclusions: Tests Proved Irresistible Spread of Radioactivity', *Life* (August 11, 1947).
102 'the lingering and insidious nature of the radioactive agent': Bradley, p. xix.
102 on the best-seller list for ten weeks: Boyer, p. 91.
102 Hair will grow back 'if the patient has not received a lethal dose': 'Feel Better Now?', *Time* (April 19, 1948).
102 'a weapon which destroys by mysterious radioactivity': Cooney, James P., 'Psychological Factors in Atomic Warfare', *American Journal of Public Health* (August 1949), pp. 969–973.
103 'a nucleus of well-trained scientists': Lapp, Ralph, *Must We Hide?* (Cambridge, Mass., 1949), p. 49.

103   'much of the danger from fallout is mental'. Gerstell, Richard, *How to Survive an Atomic Bomb* (Washington, D.C., 1950).

103   after briefly considering a site . . . near Cape Hatteras: Hacker (m.s.) VII p. 12; story of early days of the Nevada Test Site is told in Hacker (m.s.) VII, pp. 11–32.

104   more than 25 rem in four weeks: Hacker (m.s.) VII p. 24.

104   August 1, 1950, discussion of public radiation exposure: Rosenberg, pp. 26–31.

104   bombs up to 25 kilotons: Hacker (m.s.) VII p. 14.

104   weapons of up to 50 kilotons: Ibid., p. 15.

104   'make the atom routine': Ibid., pp. 16–17.

104   'the explosions wouldn't be anything more than a gag': Lang (1954), p. 68.

105   'Now that we're next door to the atom bomb': Ibid., p. 78.

105   'religious, patriotic, ordered, and simple one': Ball, pp. 53–54.

105   the shock wave that followed: Lang (1954), p. 71.

105   the night sky over San Francisco and Los Angeles: *Life* (February 12, 1951), p. 25.

105   cracked by the explosions' powerful shock waves: Lang (1954), p. 66.

106   'today it's blooming with atoms': Ibid., p. 77.

106   'There is virtually no danger': Hilgartner et al, p. 90.

106   'heavy fall-out well beyond expected ranges': Hacker (m.s.) VIII p. 39.

106   'the largest fall-out we have ever had': Ibid., p. 40.

106   Lyle Jepson won $10: Ball, p. 66.

106   'primarily of scientific interest': Hacker (m.s.) VIII p. 40.

106   'precautions must be redoubled': Ball, p. 66.

107   'gypping its public': Lang (1954), p. 78.

107   'Americans have to have their kicks': Ibid., p. 77.

107   between 250,000 and 500,000 troops: Clarfield and Wiecek, p. 205.

107   'reasons for this desired participation': Wasserman et al, p. 68.

107   a group of pigs outgrew their uniforms: Ibid., p. 71.

107   'off-site radiation for troop training purposes': *Health Impact of Low-Level Radiation*, p. 3.

107   exposure limit . . . raised to three rems: Hacker (m.s.) VIII, p. 19, p. 31; *Health Impact of Low-Level Radiation*, p. 3.

107   from one-third to two-thirds did not receive film badges: Saffer, p. 183; Noyes, Dan, O'Neill, Maureen and Weir, David. 'Operation Wigwam', *New West* (December 1, 1980).

107   The Army provided its own RadSafe: Hacker (m.s) VIII p. 29.

108   'the necessity for realistic training . . . often accompanied by serious injuries': *Health Impact of Low-Level Radiation*, p. 4.

108   'unfavourable psychological effects': Wasserman et al, p. 69.

108   'larger numbers and more serious casualties': Ibid.

108   'responsibility be adequately documented': Ibid., p. 71, and Hacker (m.s.) VIII p. 34.

109   'I would no more let my children be exposed': Bell, p. 68.

109   'The public is never told': Ibid.

109   'the subject of mutations is a touchy one': Lang, *From Hiroshima to the Moon*, p. 379.

109   'more redheaded children will be born': *Time* (November 11, 1946), p. 96.

110   'we can consider them all as bad': Ibid.

110   'the decision was made to fire on schedule': Wasserman et al, p. 72.

110   checked 250 vehicles for radioactivity: Ibid.

110   Gloria Gregerson's story: Testimony by Gloria Gregerson before US Senate Committee on Labor and Human Resources (Salt Lake City, Utah, April 8, 1982); and Ball, p. 92.

111   'Beautiful morning': Wasserman et al, p. 73.

111   radioactivity readings were . . . 6 rems per hour: Ball, p. 43.

111   warning people to stay indoors: Wasserman et al., p. 73.

111   'a little bit above normal': Ball, p. 44.

111   'the mission was to reassure residents': Ibid., p. 69.

111 'not much consolation to the sheepmen': Mazuzan and Walker, pp. 41–42.

112 'no longer had faith in the AEC': *Health Effects of Low-Level Radiation, Vol II*, p. 157.

112 'what's going to become of the world': Lang, *From Hiroshima to the Moon*, p. 379.

112 'only a single illogical and unforeseeable incident': Clarfield and Wiecek, p. 206.

112 Mike completely destroyed one island: McPhee, p. 106.

112 'winds . . . were headed for Rongelap': Johnson, pp. 12–13.

113 The account of Bravo's explosion and its effects is told in Divine, Hilgartner, Johnson, and Lapp. Dennis O'Rourke's 1986 film, *Half Life*, contains testimony from many of the Rongelap islanders who lived through the Bravo explosion.

113 'All are reported well': Lapp, p. 53.

113 The health effects of Bravo on the Rongelapese and the Brookhaven monitoring program are described in Johnson, pp. 12–16.

113 'It did not look human': Interview in *Half Life*, a film by Dennis O'Rourke.

114 'Americans are educated people': Ibid.

115 'a Red spy outfit': Divine, p. 11.

115 'far below levels which could be harmful': Ibid., p. 12.

115 'made the world fallout-conscious': Lang, *From Hiroshima to the Moon*, p. 369.

115 'talk and worry . . . is spreading': *Time* (November 22, 1954), pp. 79–81, quoted in Mazuzan, p. 43.

115 'we would have been in ignorance': Hilgartner et al., p. 100.

115 'By 1960, it was a regular flood': Wasserman, p. 64.

116 there will be no danger: Smith, Joan, *Clouds of Deceit* (London: Faber and Faber, 1985), p. 33.

116 'the implication that we know what we are talking about': Minutes of Meeting of National Advisory Committee on Radiation, November 10, 1958, p. 127.

116 'on the safe side of things': Ibid., pp. 127–8.

117 'If you ever let these numbers get out': Ibid., p. 130.

117 'certainly a confidential memo': Ibid.

117 'held off long enough': Ibid., p. 122.

117 'War with Russia is inevitable': Saffer, p. 63.

117 'it became unpatriotic': Wasserman et al, p. 93.

117 'Many persons are innocently being duped': Ibid., p. 94.

117 'The big bombs are not tested in this country': Ibid.

118 'On Your Guard': Ibid., p. 93.

118 'radioactivity may exceed safety limits': Ball, p. 79.

118 'the far greater risk': *Time* (May 6, 1957).

118 'your best action is not to be worried': Ball, p. 216.

118 'we can expect many reports': Ibid., pp. 77–78.

118 not even the JCAE knew how many bombs: Boyer, p. 305.

118 'literally squeeze information': Metzer, p. 95.

118 'selective use and release of information': Ibid..

118 'no public statement could be made'. *Health Effects of Low-Level Radiation* Vol. 1 (1979), p. 221.

119 'keep them confused': Rosenberg, p. 65.

119 'arousing public fears': Hacker (m.s.) X p. 22.

119 'over-emphasizing the effects of fallout': Ibid.

119 'the extreme care taken by the AEC': Ibid.

119 'the way to keep people fearless': Boyer, p. 305.

119 '"the limits of safety" – whatever those are': Ball, p. 68.

120 the standard was not lowered for test workers: Hacker (m.s.) VII, pp. 21–22: *Health Impacts of Low-Level Radiation*, p. 440.

120 'an unpublicized exposure of 3.9 r': Hacker (m.s.) VIII p. 16.

120   adopted the 'no-threshold' concept: Taylor (1980), p. 18.
120   'Any radiation exposure received by man': Mazuzan, p. 55.
120   'no threshold to significant injury': Hacker (m.s.) X p. 30.
120   'low-level exposure can be continued indefinitely': Clarfield and Wiecek, p. 202.
121   'the somewhat delicate public relations aspects': Hacker (m.s.) VII p. 23.
121   'safe maximum permissible dose': Hacker (m.s.) VIII p. 12.
121   'publicity considerations disregarded': Ibid.
121   'no right to subject the public': Hacker (m.s.) X p. 22.
121   'where large populations are concerned': Ibid., p. 31.
122   'If we continue to reduce the fraction we are willing to release': Ball, p. 41.
122   'killed aborning by unnecessary legislation': Mazuzan and Wiecek, p. 60.
122   'the expense to the AEC is quite appreciable': *Health Impact of Low-Level Radiation*, p. 5.
122   'reduce the cost of health protection': Ibid.
122   'create a rather bad impression': Ibid.

12 Fallout

123   Scientists outside the AEC . . . made many important discoveries: Metzer, pp. 95–96.
123   'carried all over the world': *Time* (June 10, 1946).
124   AEC's worldwide monitoring programme: US Atomic Energy Commission, Division of Biology and Medicine, *Report on Project Gabriel* (Washington, D.C., July 1954), p. 2.
124   Geiger counters began to go wild: Pringle and Spigelman, pp. 179–180.
124   The Troy fallout is discussed in Eisenbud, p. 314; Sternglass, pp. 1–5, and in Clark, Herbert. 'The Occurrence of an Unusually High-Level Radioactive Rainout in the Area of Troy, N.Y.', *Science* (May 7, 1954), pp. 619–622.
125   'The activity of drinking water': Clark, op. cit., p. 621.
125   'exceptionally high, though not hazardous': Ibid., p. 619.
125   not even attempting to regulate or measure: Metzger, pp. 85–86.
126   'there is little question': Pringle and Spigelman, p. 182.
126   'the ingestion of bone splinters': Metzger, pp. 93–94.
126   'sunshine units': Eisenbud, p. 320.
126   had ignored radioactive iodine-131: Metzger, pp. 86 and 103; Wasserman, p. 101.
126   'we were too busy chasing strontium-90': Metzger, p. 108.
127   fallout was 'not likely to be dangerous': Wasserman, p. 93.
127   'increase the existing distrust': Green, p. 57.
127   AEC meeting in February, 1955: *Health Effects of Low Level Radiation* (1979), Vol. 1, pp. 175–176.
127   'You get one group of scientists': Mazuzan, p. 57.
127   an embarrassing storm: Wasserman, pp. 94–95.
128   'Liquidating Unpopular Opinion': *Science* (October 28, 1955), p. 813.
128   'The only defensible or effective course': Mazuzan, p. 50.
128   'there is no anticipated danger': Ball, p. 40.
129   'play it down': Metzer, p. 86.
129   more than twice as many: Dennis, p. 304.
129   'civilized man would have been in trouble': Mazuzan, p. 262.
129   fallout levels reached a new peak: Mazuzan, p. 247.
129   a new furore erupted: Ibid., p. 248.
130   'more right that their critics': Ibid., p. 249.
130   'Fallout: The Silent Killer': *Saturday Evening Post* (August 29, 1959).
130   Budget officials opposed such a shift: Ibid., p. 257.
130   'It is the responsibility of the NCRP': Taylor (1979), pp. 8–168.

130 'the nation's security may demean the exposure': Clarfield and Wiecek, p. 226.
130 Among the council's six members: Metzger, p. 100.
131 recommended that the NCRP limits . . . be raised: Mazuzan, p. 260.
131 more fallout than ever before: Ibid., p. 263.
131 milk was served at every White House meal: Ibid., p. 264.
131 Dairy farmers in Utah: Ibid., p. 268.
131 Soviet tests rained iodine-131 on Minnesota: Ibid.
131 The FRC objected: Ibid., p. 269.
131 the council's guidelines did not apply to fallout: Clarfield and Wiecek, p. 228.
131 'have not caused an undue risk to health': Metzger, p. 105.
132 'the AEC said this was too restrictive': Mazuzan, p. 273.
132 existing standards were at least ten times too lax: Ibid., p. 271.
132 'the validity of the past guides'. *Health Effects of Low-Level Radiation*, p. 347.
132 more than 200 million tons of radioactive fission products: Eisenbud, p. 272.
132 every man, woman, and child on earth: Moss, p. 170.
132 4 to 5 millirems per year: Eisenbud, p. 396.
132 'The number of children and grandchildren': Wasserman et al., pp. 107–108.

13 1956 Standards

135 The BEAR committee's report: Mazuzan, pp. 44–47.
135 'any radiation . . . is unfortunate and harmful': Taylor (1979), p. 8–049.
136 a survey of AEC plant managers: Ibid.
136 'into accord with the trends of scientific opinion': Ibid.
136 'ionizing radiations will be so common': Ibid., p. 7–254.
136 'on a scale that would today be considered a major nuclear accident': Matthew Wald, 'Northwest Plutonium Plant Had Big Radioactive Emissions in 40's', *New York Times* (November 7, 1986).
136 340,000 curies of radioactive iodine: Ibid.
137 'planned experiments': Karen Dorn Steele, 'In 1949 study, Hanford allowed radioactive iodine into area air'. *Spokane Spokesman-Review* (March 6, 1986).
137 5500 curies of radioactive iodine was released: Ibid.
137 'psychologically dangerous': Taylor (1979), p. 7–124.
137 Taylor agreed, but wondered: Ibid.
137 one per cent of the occupational dose: F. D. Sowby. 'ICRP dose limits from 1934 up to 1977', *Radiological Protection Bulletin*, No. 28 (May 1979), p. 5.
137 getting the Conference to raise its recommended limit: Taylor (1979), p. 7–153; Sowby, op. cit., p. 6.
137 the ICRP also recommended: Taylor (1979), p. 7–288.
138 'rather than have two standards': Ibid., p. 8–038.
138 applied only to children: Ibid.
138 the NCRP formally followed the ICRP: Ibid., pp. 8–065 and 8–092.
138 the justification for having done so: Ibid., p. 8–199 to p. 8–204.
138 'continuing to survive and flourish': Ibid., p. 8–202.
138 'other hazards already exist': Ibid., p. 8–198A.
139 'of the order of natural background': Taylor (1979), p. 8–055.
139 *twice* as much radiation from human activities: Ibid.
139 ICRP moved closer to the NCRP's position: ICRP, 'Recommendations of the International Commission on Radiological Protection' (Pergamon Press, London, 1959).
139 'tolerable and justifiable in view of the benefit': Ibid., n. 2.
139 'may not in fact represent a proper balance': Ibid.

14 Medicine After the Bomb

140 'insufficient experimental and clinical evidence': Brecher, p. 355.

140 'When I started out': Dr James English, interview with author, July 18, 1985.

141 ringworm, acne, birthmarks: National Academy of Sciences (1977), vol. 1, p. 4.

141 more than 10,000 children in Israel: Ron, E. and Modan, B., *Journal of the National Cancer Institute*, vol. 65, no. 1 (July 1980), pp. 7–11; Modan, B., Ron, E. and Werner, A., *Therapeutic Radiology*, vol. 123 (June 1977), pp. 741–744; Shore, R. E., Albert, R. E. and Pasternack, B. S., *Archives of Environmental Health*, vol. 31, pp. 21–28.

141 'That just scared the heck out of me': English, interview with author, July 18, 1985.

141 The average dose to each child's brain: Pochin, p. 136; Gofman, p. 198.

141 'the epidemiologists were eager to get in there': Donald Thompson, telephone interview with author, June 1985.

141 six times as many cancers of the thyroid: Modan, Baruch, Elaine Ron and Abraham Werner, 'Thyroid Cancer Following Scalp Irradiation', *Radiology* (June, 1977), pp. 741–4; and Modan, B. et al. 'Radiation-induced head and neck tumours', *Lancet* 1 (1974), 277–9.

141 one of the most influential studies: Court Brown, W. M. and Doll, R. *British Medical Journal* vol. 5474 (1965), pp. 1327–1332.

141 Women who were castrated by radiation: National Academy of Sciences (1977), vol. 1, p. 5; Pochin, p. 137.

141 people whose thyroids had been treated with X-rays: National Academy of Sciences (1977), vol. 1, p. 6.

142 'people began to doubt': English, interview with author, July 18, 1985.

142 'The overwhelming majority . . . were not trained specialists': Dr Juan del Regato, telephone interview with author, July 17, 1985.

142 A survey made in 1973: Goldschmidt, H. 'Ionizing Radiation Therapy in Dermatology' *Archives of Dermatology*, vol. 111 (November 1975).

142 Hevesy found traces of radioactivity in the dinners: Tuve, M. A., 'The New Alchemy', *Radiology*, vol. 35 (August 1940), p. 180.

143 'In those days we were able to find': Brecher, p. 385.

143 more than 1,200 medical institutions: Ibid., p. 393.

144 'a serious questions as to the validity of the conclusion': National Academy of Sciences (1977), vol. 1, p. 30.

144 Detailed health records kept by the US government: Conrad, R. A., et al, 'Review of Medical Findings in a Marshallese Population Twenty-Six Years After Accidental Exposure to Radioactive Fallout', Brookhaven National Laboratory 51261, 1980.

144 'What was so horrible': Dr Francis Curry, telephone interview with author, July 1985.

145 some of the X-ray devices used: Jones and Southwood, p. 130.

145 there were other problems with the screening program: Wasserman and Solomon, p. 136.

145 'We were taking 40,000 chest X rays': Robin Jones, telephone interview with author, July 1985.

146 Curt Stern warned: Taylor (1979), p. 8–055.

146 'Our people are annually receiving more radiation': Lapp and Schubert, p. 197.

146 The so-called 'ten-day rule' was controversial: Dr Karl Morgan, telephone interview with author, October 30, 1987. Dr Raymond del Fava, chairman, American College of Radiology's Public Communications Committee, telephone interview with author, October 30, 1987.

146 'There need be no special limitations': 'The 1983 Washington Meeting of ICRP': *Radiological Protection Bulletin*, No. 57 (1984), p. 11.

146 'For a long time, people never questioned': English, interview with author, July 18, 1985.

147 'We've got to start policing ourselves': Curry, telephone interview with author, July 1985.

Notes

15 The Peaceful Atom

148 'consecrated to his life': Eisenhower's address to the United Nations General Assembly, quoted in Dennis, ed. pp. 355–356.

148 'Atomic Rays Will Cut Lumber': Pringle and Spigelman, p. 123.

149 Government-commissioned films: Ibid., p. 124.

149 featuring 'your friend, Citizen Atom': David Pensonen, 'Citizen Atom at the Park', *Sebastopol Times* (September 27, 1962).

149 Atomic Energy Merit Badge: Hilgartner et al., p. 74.

149 Glenn Seaborg's predictions: Glenn Seaborg and William Corliss, *Man and The Atom* (New York: E. P. Dutton and Co., Inc., 1971), pp. 188–9 and 174.

149 'electrical energy too cheap to meter': Ford, p. 50.

149 Project Plowshare is described in Lewis, pp. 172–198.

150 Gasbuggy: Lewis, Ibid.; Clarfield and Wiecek, pp. 366.

150 'The First Dinners Are Cooked!': J. Cutler, 'British Nuclear Foul-up Limited', *New Statesman* (November 18, 1983).

150 The story of the Windscale fire is told in Eisenbud, pp. 351–357; Clarfield and Wiecek, p. 349; and Wasserman et al., pp. 170–171.

151 the Windscale fire will have caused about 33 deaths: R. Edwards and J. Cutler, 'Overshadowed Only By Chernobyl', *The Independent* (October 5, 1987).

151 The story of the Fermi reactor is told in John Fuller. *We Almost Lost Detroit* (New York: Berkley Books, 1984); Clarfield and Wiecek, pp. 193–197; and in Wasserman et al., pp. 208–209.

151 nuclear power in the United States proceeded slowly at first: Atomic Industrial Forum, *Historical Profile of U.S. Nuclear Power Development* (1985); Wasserman et al., pp. 283–294; and Ford, p. 62.

151 less of the nation's energy than did firewood: Ford, p. 179.

151 'We didn't know much about radiation': David Pensonen, telephone interview with author, January 27, 1987.

152 Pacific Gas & Electric dropped its plans: Dennis, ed., p. 437.

152 An AEC study in 1956: Ford, p. 45.

152 'frighten the country and the Congress to death': Ibid., p. 46.

16 Confusion and Acrimony

153 a study published in the *British Medical Journal*: A. Stewart, et al., 'A Survey of Childhood Malignancies', *British Medical Journal* (June 28, 1958).

154 'The antagonism was from the medical profession': Dr Alice Stewart, interview with the author, December 5, 1984.

154 Stewart's work was confirmed beyond doubt: B. MacMahon, 'Prenatal X-ray Exposure and Childhood Cancer', *Journal of the National Cancer Institute* 28 (1962).

154 Gofman had close ties to . . . Seaborg: Dr John Gofman, interviews with author, August 28, 1984 and January 29, 1987.

154 'twenty times greater than was being said': Ibid.

155 The story of the Gofman and Tamplin furore is taken from a number of sources, including *Science* (August 28, 1970), pp. 838–843; Lewis, pp. 81–108; Wasserman et al., pp. 208–215; and the author's interviews with John Gofman on August 28, 1984 and January 29, 1987.

155 Sternglass submitted a paper to *Science*: Sternglass, pp. 93–118.

156 'even more exaggerated claims by the AEC': Lewis, p. 82.

156 'decided to use this talk': Gofman, interview, January 29, 1987.

156 'an excess of 32,000 cases of fatal cancer': Gofman and Tamplin, p. 79.
156 'To continue the present guidelines is absolute folly': M. Schwartz, 'Safe Radiation Limit Too High': *San Francisco Chronicle* (October 30, 1969).
156 'Gofman and Tamplin do not make a case for revision': Lewis, p. 92.
156 that the two men were 'incompetent': Gofman, interview, January 29, 1987.
157 'I am not in agreement with Gofman and Tamplin': Taylor (1979), p. 10–390.
157 'the beginning of the end': Gofman, interview January 29, 1987.
157 'a scientific whorehouse': *Science*, op. cit., p. 842.
157 'partially succeeded in silencing our presentation': Ibid., p. 840.
158 threatened to delete a quarter of a million dollars: Gofman, interview, August 2, 1984.
158 in Great Britain, 200,000 people die of cancer: American Cancer Society, 'Cancer Facts and Figures – 1988' (New York City), p. 31.
158 the United States has about 500,000 cancer deaths: Ibid.
158 necessary to study ten million exposed people: Hendee, p. 206.
159 Jay Gould study: *The Independent* (January 4, 1988). *The Economist* (January 30, 1988), p. 67.
160 'I have been repeatedly urged by AEC': *Effect of Radiation on Human Health* (vol. 1), p. 525.
160 'aside from a certain political usefulness': Janet Raloff, 'Radiation Politics', *Science News* (February 10, 1979), pp. 93–95.
160 'much of the motivation for starting this study': Ibid.
161 to fight workman's compensation claims: *Bulletin of Atomic Scientists* (February 1980), p. 62.
161 the statement they wished him to release: *Effect of Radiation on Human Health* (vol. 1), p. 559.
161 the AEC would not renew his contract: *op. cit.*, p. 567.
161 'the message is clear': Wasserman, p. 143.
161 approximately six per cent more deaths: Mancuso, T. F., Stewart, A., and Kneale, G., 'Radiation Exposure of Hanford Workers Dying From Cancer and Other Causes', *Health Physics* 33 (November 1977).
161 'They urged us not to publish': Wasserman, p. 143.
162 special exemption from federal conflict-of-interest statutes! *Bulletin of Atomic Scientists* (February, 1980), p. 62.
164 re-evaluation of the 1980 BEIR risk estimates: Preston, Dale and Pierce, D. A., 'The Effect of Changes in Dosimetry on Cancer Mortality Risk Estimates in the Atomic Bomb Survivors', *Radiation Effects Research Foundation Technical Report 9–87*.
164 Edward Radford . . . also gives an estimate of risk: Jones and Southwood, p. 157.
164 Another RERF study: Shimuzu, Y. et al., 'Life Span Study Report, Part I', RERF Technical Report 12–87; 'Life Span Study Report, Part II', RERF Technical Report 5–88.
164 'an illusion that there is a radiation effects controversy': Arthur Tamplin. 'Issues in the Radiation Controversy', *Bulletin of the Atomic Scientists* (September, 1971), p. 25.
164 'do not yield results which are very far from present risk numbers', Taylor (1979), p. 10–390.
165 'we are coming to a relatively consistent estimate': Jones and Southwood, p. 158.
165 'will probably increase, perhaps substantially': NCRP, Report No., 91 (1987), p. 23.

17 A House Divided

167 at least sixteen federal agencies: Taylor, L. S., *Health Physics* vol. 39 (December 1980), pp. 861–2.
168 'Maybe the facts should be laid out': Taylor (1979), p. 8–475.
168 'because we don't have the information': Taylor, interview with author, September 8, 1984.
168 'a certain amount of incest': John Dunster, interview with author, October 7, 1985.
169 about $150,000 a year: Ibid.
169 at least 1.5 million dollars: Ibid.

169 'almost 40 research laboratories': *Capsule Review of DOE Research and Development and Field Facilities* (September, 1986) US Department of Energy, p. xvii.
170 University of California, for example: Ibid., pp. 33–36, and pp. 55–57.
170 In 1985, the DoE spent $140 million: 'Statement of Alvin W. Trivelpiece, Director of Energy Research, DoE, before the Subcommittee on Natural Resources, Agriculture and Environment of the House Science and Technology Committee, 26 March, 1985', p. 6.
170 'There are very few people in radiation protection': Richard Guimond, interview with author, September 11, 1984.
170 names more than thirty researchers in ten countries: Martin, Brian, *Science and Public Policy* vol. 13, number 6, (December 1986), pp. 312–320.
170 'If you mind your own business and keep quiet': Dr Rosalie Bertell, interview with author, May 2, 1984.
170 'It's just an automatic thing': Dr Robert Blackith, interview with author, October 16, 1984.
171 'We react like mechanical puppets': Taylor (1979), p. 10–389.
171 'a small handful of pseudo-scientists': Taylor, Lauriston, 'What the Public Is Told and What It Should Know About Radiation Hazards', *Health Physics Society's Newsletter*, vol. XI, number 8 (August, 1983).
171 'a few hungry charlatans': Ibid.
171 'one-time scientists': Taylor, Lauriston, 'Some Nonscientific Influences on Radiation Protection Standards and Practice', *Health Physics* Vol. 39 (December, 1980), p. 869.
171 'this group of rather sloppy people': Taylor, interview with author, August 8, 1984.
171 'not part of the radiation protection community': Taylor, Ibid.
171 'terror-mongers': Taylor, Lauriston, 'Let's Keep Our Sense of Humour in Dealing with Radiation Hazards', *Perspectives in Biology and Medicine*, Volume 23, number 3 (Spring 1980), p. 334.
171 'the tragedy of this': Ibid., p. 326.
171 'frivolous charges and claims': Taylor, Lauriston, 'Radiation Protection Standards: What They Are and What They Are Not', Paper presented at Eighth Life Sciences Symposium, Los Alamos Scientific Laboratory (October 9, 1980), p. 25.
172 'They represent themselves': Dr Ralph Lapp, interview with author, June 17, 1987.
172 'I was public enemy number one': Ibid.
172 'claim it's classified': Ibid.
172 'In those days you were alone': Ibid.
173 'Karl Morgan is a gentleman': Ibid.
173 'Gofman's estimates are simply weird': Ibid.
173 'Mancuso is a politician': Ibid.
173 'tools of the anti-nukes': Dr Thomas Cochran, interview with author, September 11, 1984.
173 'Morgan, Nader, Fonda, etc.': Taylor, Lauriston, 'Radiation Protection Standards: What They Are and What They Are Not', Paper presented at Eighth Life Sciences Symposium, Los Alamos Scientific Laboratory (October 9, 1980), p. 12.
173 Gofman supports a nuclear defense: Gofman, interview with author, July 28, 1984.
173 'I'm for the safe development of nuclear power': Morgan, interview with author, August 13, 1984.
173 'A lot of people are over-worried': Gofman, interview with author, June 2, 1987.
173 'prudence seems to dictate silence': Abelson, Philip, quoted in Lapp, Ralph, *The New Priesthood* (New York: Harper & Row, 1965), p. 30.
173 'It takes a very stubborn individual': Dr Victor Gilinsky, interview with author, March 26, 1986.
174 'I learned then what it was like to be in a minority': Stewart, interview with author, December 5, 1984.
174 'New ideas take a generation': Ibid.
174 'no reputation to lose': Taylor, interview with author, August 8, 1984.

174 'gone round the bend': Sowby, telephone interview with author, July 13, 1984.

174 'It's a potential motive': Taylor, interview with author, August 8, 1984.

175 'Dr Morgan's testimony is stricken from this case': Kelly, Patrick, 'Memorandum and Decision' in *Johnson* v. *U.S.* (1984), p. 88.

175 'impressive' and 'brilliant': Ibid., pp. 109–118.

175 'he's close to a saint': Gofman, interview with author, July 28, 1984.

175 'if your objective is wealth': Cochran, interview with author, September 11, 1984.

175 'friends that do things you hate': Morgan, interview with author, August 13, 1984.

175 'biased in their views': Ibid.

175 'I don't have to please the boss': Ibid.

175 'A large fraction of members of ICRP': Karl Morgan, letter to author, August 13, 1984.

176 'In spite of its usefulness in the past': Green, Patrick, *The Record of the International Commission for Radiological Protection* (London: Friends of the Earth, 1987), p. 18.

176 the public's concern ... is 'out of proportion': National Radiological Protection Board, *Radiological Protection Bulletin* (April 1986), p. 3.

176 'Our society expects a higher standard of protection': Dunster, John, 'Are we too frightened of radiation?', *New Scientist* (October 13, 1983).

176 'Anxiety or even phobia': Ibid.

176 'more than we have spent on all other environmental measures': William Waldegrave, 'Speech to the Natural Environment Resource Council' (April 1, 1987), p. 6.

176 'are we being conscientious or just plain silly?': National Radiological Protection Board, *Radiological Protection Bulletin* (November, 1985), p. 3.

176 an unofficial rule of thumb: Eisenbud, p. 43.

177 it costs of only $40,000 to save a life with smoke detectors, and only $3000 to save a life with seat belts: Eisenbud, p. 400.

177 to find ways to overcome the public's 'nuclear phobia': Kurtz, Howard, 'U.S. Probes Fear of Nuclear Power', *Washington Post* (October 30, 1984).

177 'In short, we the general public, are irrational and uninformed': Allman, William, 'The Risk Game', *Science 85*, reprinted in the *San Francisco Chronicle* (October 13, 1985).

177 'it is only in the use of medical procedures': Taylor, Lauriston (1980), p. 855.

177 'People hate to contemplate dying in strange ways': Norman C. Rasmussen, quoted in 'Technology Missteps: Atomic Difference', *The New York Times* (May 5, 1986).

177 'acceptable political or economic reasons to warrant it': Taylor, Lauriston (1980), p. 867.

178 'Our fears may tell us a lot': Allman, op. cit.

18 1977 Standards

179 no amount of radiation was absolutely safe: Taylor (1980), p. 18.

179 1-in-10,000 risk of accidental death per year is acceptable: ICRP-26, pp. 20–21.

180 accidental death rate for the manufacturing industry in 1986: data from National Safety Council, Washington, D.C., 1986.

180 'about as dangerous as the public will accept': Sowby, interview with author, September 12, 1985.

180 an annual accidental death rate of about 1 in 10,000: National Safety Council, Washington, D.C., 1986 data.

180 a 6-in-10,000 chance of dying: The ICRP estimates that one million person rems will result in 125 cancer deaths among the exposed population and 80 serious genetic defects in succeeding generations. Thus one rem of exposure carries a 1-in-8,000 risk of cancer death and a 1-in-12,500 risk of serious genetic defects. A five-rem exposure would increase those risks by a factor of five, producing a 6-in-10,000 chance of fatal cancer and a 4-in-10,000 chance of serious genetic defects. The ICRP does not include non-fatal

cancers in its risk estimates, but most authorities say that they are from 1.5 to 2 times as common as fatal cancers. So a five-rem exposure would entail at least a 9-in-10,000 chance of developing a non-fatal cancer.

180 'It wouldn't really be tolerable': Sowby, interview with author, September 12, 1985.

180 97 per cent of all US radiation workers: EPA, 'Occupational Exposure to Ionizing Radiation to the United States, 1960–1985' (September, 1984), p. 44; EPA, 'Federal Radiation Protection Guidance for Occupational Exposure: Response to Comments' (January, 1986).

180 EPA data on average occupational doses: EPA, op. cit., pp. 40 and 42.

181 'you could not in effect run anything': David Harward, interview with author, May 13, 1985.

181 'a 5 rem annual limit is needed': EPA 1986, op. cit., p. 52.

181 'hiring or training qualified replacement workers is difficult or impossible': Ibid.

181 reducing the 5-rem limit could cause a shortage of certain professionals: Ibid.

181 would cost hospitals several hundred million dollars a year: Ibid.

181 'They'd have to work a lot harder at it': Gilinsky, interview with author March 26, 1986.

181 corroded tubes in their steam generators: *Wall Street Journal* (October 12, 1983).

182 would cost the average nuclear power plant more than $6 million: Atomic Industrial Forum, *Study of the Effects of Reduced Occupational Radiation Exposure Limits on the Nuclear Power Industry* (January 1980), p. 120.

182 eliminate 2.6 million person rems: US Nuclear Regulatory Commission, Office of Standards Development, 'Instruction Concerning Risk from Occupational Radiation Exposure' (May 1980), p. 21.

182 an additional $1.5 millign each year: Atomic Industrial Forum, op. cit., p. 120.

182 almost 30,000 new radiation workers: EPA 1986, op. cit., pp. 52–53.

182 'severely restrict specialty workers': Ibid.

182 suggested that the limit be lowered to 1 rem or less: Roger Kasperson and John Lundblad, *Environment* (December 1982), pp. 16 and 37.

182 EPA has considered and rejected that idea: EPA 1986, op. cit., p. 69.

182 'very bad to pick out a class of workers': Dave Harward, interview with author, May 13, 1985.

182 'The commission didn't adopt a new number': Ibid.

182 'all exposures shall be kept as low as reasonably achievable': ICRP-26, p. 3.

183 'it's that tenth that is the acceptable limit': Sowby, interview with author, September 12, 1985.

183 'ALARA says, If it doesn't cost an unreasonable amount': Dr Michael Thorne, interview with author, September 12, 1985.

183 'It's a matter of national and local judgment': Ibid.

183 Different US Government agencies have valued a human life: *The New York Times* June 26, 1985, p. 1.

184 'Government scientists are in chains': Sir Kelvin Spencer, Letter to the editor, *The Guardian* (July 22, 1983).

184 'another yank on the ratchet', Taylor, interview with author, September 8, 1984.

185 'There's an awful lot of people who feel that there's just no justification', William Beckner, interview with author, June 16, 1987.

185 'You can just forget it completely': Ibid.

185 'a bottom line for ALARA': Ibid.

185 Revised annual exposure limits for various organs of the body: ICRP-26, p. 21.

186 For about 170 radionuclides, the limits were raised: *Federal Register*, Thursday January 9, 1986, Part II, Nuclear Regulatory Commission, 10 CFR Parts 19 et al., pp. 1119–1120; Natural Resources Defense Council, *Comments on the Proposed Standards for Protection Against Radiation* (1986), p. 6; Robert Alvarez, interview with author, August 7, 1984.

186 'undoubtedly will be raised by our licensees': Robert Alexander, interview with author, September 10, 1984.

186 'what about India, what about the Philippines?': Robert Alvarez, interview with author, August 7, 1984.

186 'I was a critic and I still am': Alexander, interview with author, September 10, 1984.

186 the NRC has proposed: *Federal Register*, Thursday January 9, 1986, Part II, Nuclear Regulatory Commission, 10 CFR Parts 19 et al.

186 'it gives the whole thing more respectability': Gilinsky, interview with author, March 26, 1986.

187 'would be likely to be acceptable': ICRP-26, p. 23.

187 'the situation might still be justifiable': Ibid., p. 24.

187 proposing to insist that these facilities not expose their neighbours to more than 100 millirems per year: 'Part II. Nuclear Regulatory Commission. 10 CFR Parts 19 et al.', *Federal Register*, January 9, 1986, p. 1133.

187 'One of the main reasons why we can't reduce the limit further': Hal Peterson, interview with author, September 8, 1987.

188 'it would be a nightmare to regulate': Ibid.

188 'a cost-effectiveness analysis, not a health analysis': Ibid.

188 'literally in the 1940s and early fifties': Robert Alvarez, interview with author, August 7, 1984.

188 'National security reasons do not allow the permanent shutdown': *San Francisco Chronicle* (December 13, 1986), p. 11

189 'an increase in the average dose to members of the public': ICRP-26, p. 25.

189 'the principal limit is 100 millirems in a year': 'Statement from the 1985 Paris meeting of the International Commission on Radiological Protection', *Annals of Physics, Medicine, and Biology* (1985), vol. 30, No. 8, p. 863.

19 Natural Radiation

193 correlations between background radiation and childhood leukemia: *New Scientist* (October 23, 1986), p. 15.

193 about one per cent of all fatal cancers: Pochin, p. 174.

194 A 150 pound person contains about 140 milligrams of radioactive potassium: Eisenbud, p. 149.

194 high levels of radioactive lead-210 and polonium-210 in [Laplanders and Eskimos]: Ibid., pp. 147–8.

194 drinking water supplies in several sizable towns . . . contain radium: Ibid., p. 133.

195 'the potential problem of radon released from domestic waters': Kathren, p. 67.

195 the average person receives a dose of about 100 millirems a year: 'Interim Guidance on the Implications of Recent Revisions of Risk Estimates and the ICRP 1987 Como Statement', National Radiological Protection Board, NRPB-GS9 (November, 1987), p. 6.

195 People living in Denver: Eisenbud, p. 160.

195 The dose at the top of Mount Everest: Pochin, op. cit., p. 62.

195 emitting 17,500 millirems per year: United Nations Environment Program, *Radiation: Doses, Effects, Risks*, UNEP (1985), p. 16.

195 from 800 to 1500 millirems per year: Ibid.

196 they can produce X-ray photographs of themselves: Kathren, p. 78.

196 an ordinary aeroplane across the Atlantic: Kathren, p. 78.

196 the same trip on Concorde: Ibid.

196 more radiation than workers are allowed to get: Eisenbud, p. 163.

196 could eventually double the amount of radium and uranium naturally present: Ibid., p. 151.

196   30 million person-rems of radiation: UNEP op. cit., p. 46.

196   3,000 curies of radioactive potassium to US farmlands: Eisenbud, p. 149.

197   burning coal in a power station produces a collective dose-commitment of 200 person-rems: UNEP, p. 24.

197   the collective dose commitment from generating one gigawatt-year of electricity with nuclear power: Ibid., pp. 41–2.

197   the average radon dose is 80 millirems per year: NRPB op. cit., p. 6.

198   Watras' annual dose from radon was estimated to be 40 rems a year: Erik Eckholm, 'Radon: Threat Is Real, but Scientists Argue Over Its Severity', *New York Times* (September 2, 1986), p. 19; Geoffrey Lean, 'Radon: second largest cause of lung cancer', *The Observer* (July 13, 1986).

198   from 5,000 to 20,000 of the 136,000 fatal lung cancers Americans suffer every year: Reeves, Pamela, 'Homes may get radon test check', *San Francisco Examiner* (August 19, 1987).

198   increase his or her risk of dying from lung cancer: Eckholm, op. cit.

198   Non-smokers would at least double their normal . . . risk: Ibid.

198   the average American house was believed to contain roughly one picocurie of radon per litre: Ibid.

198   as many as 12 per cent of the country's 70 million homes: Ibid.

198   radon doses in one million American homes: *New Scientist* (February 5, 1987), p. 34.

20   Multiple Exposures

201   more than 500 licensed nuclear facilities: State of California, Department of Health Services, Radiological Health Branch, September 16, 1987.

201   'If you take any one source of exposure': Stanton, Kathleen, 'Uranium: In the atom's shadow', *The Arizona Republic* reprint of series that ran December 5–12, 1982, p. 12.

201   fifty US commercial reactors emitted 40 million curies: Wasserman, Harvey, 'Three Mile Island Did It', *Harrowsmith* (May–June, 1987), p. 55.

201   the public's dose from nuclear power operations: Lean, p. 47.

201   the collective dose will rise to 1,000,000 person-rems: Ibid.

202   the nuclear fuel cycle produces about 400,000 person-rems: Ibid., p. 42.

202   23,000 in the United States alone: Greg Cook, Nuclear Regulatory Commission, telephone interview with author, September 11, 1987.

203   33 citations for careless handling of radioactive materials: Davidson, Keay, 'UCSF may get stiff fine for radiation violations', *San Francisco Examiner* (September 2, 1987).

203   they have twice as much radioactive iodine: Lean, Geoffrey, 'Stockbroker belt worse than N-plant', *The Observer* (February 8, 1987).

204   191 million tons of tailings at 52 sites: Crawford, Mark, 'Tailings: A $4-Billion Problem', *Science* (August 4, 1985), p. 537.

204   give the nearest residents a 1-in-1000 chance of premature death: Stan Lichtman, interview with author, September, 1987.

205   'there were very few people living near the piles': Ibid.

205   'Why should those people be less protected?': Linda Taylor, interview with author, April 22, 1985.

205   'just moved their families out again': Greg Hoch, interview with author, April 29, 1985.

206   'at least one case of cancer or a cancer-related death': Collier, Sylvia, 'The Killing Field', *The Observer Sunday Magazine* (March 10, 1985), p. 14.

207   eight millirems per year to the reproductive organs of the wearer: Kathren, p. 79.

207   8.4 million radium timepieces in use: Ibid.

207 total dose to all workers: Environmental Protection Agency, 'Occupational Exposure to Ionizing Radiation in the United States' (September 1984), p. 40.

208 recommended that such glazes be banned entirely: US Nuclear Regulatory Commission, 'Environmental Assessment of Consumer Products Containing Radioactive Material' (October 1980), p. 85.

208 as much as 4000 millirems per year: Poch, p. 25.

208 'few dentists were aware of the fact': NRC, op. cit., p. 4–22.

208 each crown gives a dose to the mouth of approximately 1000 millirems: Ibid., p. 4–14.

208 'simply cannot be established with the presently known data': Ibid., p. 4–22.

209 a million or more of these kits: Hoaward Kartstein, telephone interview with author, September 12, 1987.

210 'The suggested way to dispose of the kits is down the drain': Gerald Wong, telephone interview with author, September 10, 1987.

210 General Electric company had to recall 90,000 sets: Brodeur, Paul, *The New Yorker* (December 13, 1967), p. 96.

210 average dose to the gonads: NAS 1980, p. 63.

210 3 billionths of a rem per hour: *New Scientist* (October 9, 1986), p. 66.

210 has been doubling every 15 years: EPA, op. cit., p. 54.

210 one and a half million radiation workers in the United States: Ibid.

211 the collective dose for radiation workers rose: Ibid., p. 59.

211 average dose to American students: Ibid., p. 42.

211 managed to cut individual doses: Brooks, p. 26.

211 fifty per cent more radiation exposure to workers: Ibid.

211 in the United States is 1,500 millirems: Ibid.

211 400 and 200 millirems, respectively: Alexander, Robert, *Health Physics Newsletter* (July 1986), p. 6.

212 1,500 welders, each working for only 15 minutes: Franklin, Ben. A., 'Nuclear Plants Hiring Stand-Ins To Spare Aides Radiation Risks', *New York Times* (July 16, 1979); Grossman, Richard and Daneker, Gail, *Energy, Jobs, and the Economy* (Boston: Alyson Publications, MA, 1979), p. 82.

212 temporary workers at nuclear power plants increased by forty times: Brooks, Barbara, 'Occupational Radiation Exposure at Commercial Nuclear Power Reactors 1982', NRC (December 1983), p. 32.

212 almost half of all nuclear power plant workers: Atomic Industrial Forum, 'Characterization of the Temporary Radiation Work Force at U.S. Nuclear Power Plants' (May 1984), p. iii.

212 'the thousand or more workers that might be brought in': Ibid., p. 7.

213 'We find that training often reduces the impression of hazard': Melville, Mary, 'The Temporary Worker in the Nuclear Power Industry', Center for Technology, Environment, and Development (Clark University, Worcester Massachusetts, 1981).

213 more than 5,264 worked at two or more different stations: Brooks, op. cit., 1983, p. 32.

213 whole-body doses of 5 rems or more: Ibid., p. 34.

213 grew by a factor of forty: Ibid., p. 31.

213 'There's a lot of worry about this': Lauriston Taylor, interview with author, August 8, 1984.

213 'workers in some instances do not divulge previous work': Melinda Renner, interview with author, May 13, 1985.

213 'he can do that without breaking any regulation' Robert Alexander, interview with author, September 10, 1984.

21 Keeping Track

214 five curies of radioactive material down the drain: Hal Peterson, interview with author, September 8, 1987.

215 'If everyone who can do it, does it': Ibid.

215 'periodically, but not frequently, monitor the water': Ibid.

215 residents of industrialized countries receive an average dose: NCRP Scientific Committee 48, 1987.

216 'the lower end of the range': Franke, B., E. Kruger, B., Steinhilber-Schwab, H. van de Sand, D. Teufel, 'Radioactive Exposure to the Public From Radioactive Emissions of Nuclear Power Stations', Institute of Energy and the Environment, University of Heidelberg, n.d.

217 'validated in the field': Eisenbud, p. 51.

217 Scientists found: J. L. Heffter, J. F. Schubert and G. A. Mead, 'Atlantic Coast Unique Regional Atmospheric Tracer Experiment (Accurate),' NOAA Technical Manual, ERL-ARL-130 (October 1984), Silver Springs, MD.

217 'we didn't find that': William Lawless, telephone conversation with author, June 6, 1988.

218 'a choice that we knew from experienced judgment was right': Lord Hinton's speech quoted in *The Ecologist Digest*, Vol. 11, No. 5 (1981).

219 'quite certain that no accumulation of radioactivity can occur': Proceedings of Session D-19, Atoms for Peace Conference (September 11, 1958), p. 623.

219 'use radioactivity and discharge it and find out what happens to it': Ibid., p. 624.

219 more than a quarter of a ton of plutonium: Jason Adkins, *Multinational Monitor* (July 1984), p. 15.

219 Seventy-five per cent of the total radiation dose: W. C. Camplin and M. Broomfield, 'Collective Dose to the European Community from Nuclear Industry Effluents Discharged in 1978': National Radiological Protection Board (1983).

219 These models have turned out to have several shortcomings: John Dunster, interview with author, October 7, 1985.

220 'We just can't find enough radiation': Ibid.

220 40 times higher than had been officially recorded: *The Observer* (February 23, 1986), p. 11.

220 'no rational cost effective basis for doing so': British Nuclear Fuels Ltd., *Highlights For 1983*, pp. 1–2.

220 'it will seldom be possible to ensure that no individual exceeds the dose limit': International Commission on Radiological Protection, *Recommendations of the ICRP* (1965), pp. 8–9.

221 'You can quibble about the accuracy': Robert Alvarez, interview with author, August 7, 1984.

221 'not performing with an acceptable degree of consistency and accuracy': *The Progressive* (August, 1981), p. 18; NRC advisory notice (March 28), 1981.

221 exposures as high as 7 rems: *Wall Street Journal* (September 4, 1980).

222 The other 7,000 do not have to report their exposure data: Barbara Brooks, telephone interview with author, March 31, 1987.

222 the average exposure of industrial radiographers is around one rem: Ibid.

222 'they may get doses as large as radiographers': Ibid.

222 'you could double the listed exposure': Ibid.

222 errors in monitoring and record-keeping: D. W. Moeller and L. C. Sunt, 'Personnel Overexposures at Commercial Nuclear Power Plants', *Nuclear Safety*, Vol. 22, No. 4 (July–August 1981), pp. 501–502.

222 half the power plants inspected had inadequate training and retraining programs: Ibid., p. 504.

223 'serious weaknesses in the control of personnel contamination': Ibid.

223 approximately 30,000 health physicists worldwide: John Poston, telephone interview with author, September 17, 1987.

223 training is mostly on-the-job: Ibid.

223 Most health physicists who have attended college: D. W. Moeller, *Health Physics Journal* (July 1971), quoted in Bertell, pp. 60–61.

223 'there should be no surprise and no recriminations about the consequences': John Tolan, *Health Physics Newsletter* (July 1986), p. 2.

22 Unnecessary Exposures

224 More than 90 per cent of all non-background radiation: Kathren, p. 90; Daryl Bohning, in Jack Dennis, ed., p. 244.

224 'the medical profession is a greater problem than the nuclear industry': Morgan, interview with author, August 13, 1984.

224 almost half the total radiation dose comes from medicine: Kathren, p. 90; Eisenbud, p. 396; Committee on the Biological Effects of Ionizing Radiations, p. 66.

224 3,670 deaths and an unknown amount of genetic damage: Laws, 1983, p. xv.

224 16,000 deaths in the United States each year: interview with author, August 13, 1984.

224 58 million medical X-ray exams were performed in hospitals: Ralph Bunge, chief of Trends Analysis Branch of US Bureau of Radiological Health, telephone interview with author, July 1 1985, statistics from BRH's unpublished 'radiation experience data'.

224 in 1970 the number had grown to 82 million: Ibid.

224 by 1980 it was 130 million: Ibid.

224 X-rays taken outside hospital: Ibid.

225 the average person's annual dose from diagnostic X-rays: Dr Fred Mettler, telephone interview with author, October 30, 1987.

225 In China, for example, 98 per cent of all X-ray diagnosis: Ibid.

225 90 per cent of the work done by plastic surgeons in China: Ibid.

225 60 per cent of the X-ray machines in Bangladesh: Ibid.

225 testing the white blood cell count of the technicians: Ibid.

225 UNSCEAR estimates: Ibid.

226 the use of radiopharmaceuticals increased by five times: Bohning, op. cit., p. 244.

226 radiopharmaceuticals were used approximately 11 million times: Eisenbud, p. 238.

226 a chest X-ray taken using modern equipment: US Department of Health and Human Services, *National Evaluation of X-Ray Trends Tabulations: Representative Sample Data* (November, 1984). p.C.

226 an examination of the lower gastro-intestinal tract: Laws (1983). Appendix C.

226 A complete abdominal exam: Laws (1983), p. 5.

226 median dose to the skin from a chest X ray: US Department of Health and Human Services (1984), op. cit., p.C.

226 skin dose for an abdominal exam: Ibid.

226 Skin doses from a dental bitewing: Ibid., p.D.

226 30 per cent of all X-ray machines failed to comply: Charles, Charles and Michael Hudson, *The Silent Intruder* (New York: Pan Books, 1982), p. 61.

227 average exposures in Illinois fell: US Department of Health and Human Services, *Patient Radiation Exposure in Diagnostic Radiology Examinations* (August 1983), p. 8.

227 lowered the average skin dose for a single dental X-ray: US Department of Health, Education and Welfare (FDA), *Population Exposure to X-rays, U.S. 1970* (November 1973), p. 55; US Department of Health and Human Services, *Nationwide Evaluation of X-Ray Trends (NEXT) 1974–1981* (April 1984), p. 10.

227 'mean exposure levels for medical examinations remained relatively stable': US Depart-

Notes

ment of Health and Human Services, *Nationwide Evaluation of X-Ray Trends (NEXT) 1974–1981* (April 1984), p. 10.

227 The mean dose to internal organs from chest X-rays actually increased: Ibid., pp. 43–44.

227 'have no recognised credentials': House Committee on Interstate and Foreign Commerce, Subcommittee on Oversight and Investigations, *Unnecessary Exposure to Radiation from Medical and Dental X-rays* 96th Congress (July 29–31, 1979).

227 taken by a completely uncredentialed person: Ibid., p. 93.

227 a landscape gardener who X-rays tree trunks: Ibid., p. 50.

228 'all those technical things that you forget': interview with author, June 26, 1985.

228 'Dentists should be far more knowledgeable': telephone interview with author, July 1, 1985.

228 51 per cent of California's dental practices: Donald Bunn, telephone interview with author, June 20, 1985.

228 'A good percentage of the doctors who take our test fail': Bramer, Ibid.

228 only about 80 per cent of doctors eventually pass: House Committee on Interstate and Foreign Commerce, Subcommittee on Oversight and Investigations, *Unnecessary Exposure to Radiation from Medical and Dental X-rays* 96th Congress (July 29–31, 1979).

228 The most recent survey of X-ray equipment and techniques: Ibid., p. 65.

228 'things have improved since then': Donald Bunn, telephone interview with author, October 21, 1987.

228 30 X-ray inspectors: Ibid.

228 a single X-ray from those machines: Donald Bunn, telephone interview with author, June 20, 1985.

228 The same procedure carried out by a 100 kilovolt machine: Ibid.

228 'dictate to medicine': Bramer, interview cited.

229 'there's not much training in the schools': Bunn, 1985 interview.

229 'Developing properly is really the critical thing': Bramer interview.

229 63 per cent of those without credentials: Wasserman and Solomon, p. 129.

229 survey of X-ray trends between 1974 and 1981: US Department of Health and Human Services, *Nationwide Evaluation of X-Ray Trends (NEXT) 1974–1981* (April 1984), p. 42.

230 reduce the radiation dose to male sex organs by 75 per cent: Laws (1983), p. 17.

230 one dentist in four shields patients: House Committee on Interstate and Foreign Commerce, Subcommittee on Oversight and Investigations, *Unnecessary Exposure to Radiation from Medical and Dental X-rays* 96th Congress (July 29–31, 1979), p. 145.

230 more than one in ten X-ray exams are worthless: Bureau of Radiological Health, 'Background Paper on Unnecessary X-Ray Examinations' (June 1978), p. 70.

230 half the films used to diagnose black lung disease: Laws (1977), p. 112.

230 one in five X-ray exams *is* unnecessary: Bureau of Radiological Health (June 1978), op. cit., p. 70.

231 'anxiety in patients or in doctors': Laws (1983), p. 11.

231 'An excessive number of X-rays may be taken': Laws (1977), p. 84.

231 Seventy-five per cent of the doctors polled: Wasserman and Solomon, p. 137.

231 'If a tennis player suffers elbow pain': Ibid., p. 136.

231 'you've got to do enough business': James English, interview with author, July 18, 1985.

231 twice as likely to be X-rayed: Herbert Abrams, *New England Journal of Medicine* (May 24, 1979), p. 1215.

231 more than 18 million unnecessary X-rays: Bureau of Radiological Health, 'Background Paper on Unnecessary X-Ray Examinations' (June 1978), p. 69.

232 35 million routine chest X-rays. Eugene Rabin. 'Second Opinion', *San Francisco Examiner*, July 16, 1987.

232 the leading cause of cancer death in American women: Laws, p. 13.

232 'waiting lists for mammograms': Wasserman and Solomon, p. 132.

277

232 1,800 cases of breast cancer were detected: Ibid.
232 produce five cancers for every one detected: Poch, p. 125.
233 women younger than 50 should not be mammogrammed: Laws (1977), p. 66.
233 a recent seminar at the University of California: Proceedings of a seminar on 'Nuclear Technology, Society and Health' (October 16, 1986), p. 12.
233 many doctors failed to ask women: Wasserman and Solomon, p. 130.
233 266,000 pregnant women in the United States were X-rayed for that reason, Ralph Bunge and Carol Herman, *Radiology* (May 1987), p. 571.
234 800,000 unnecessary emergency room skull X-rays: Bureau of Radiological Health, 'Background Paper on Unnecessary X-Ray Examinations' (June 1978), p. 69.
234 the cost of unnecessary skull X-rays: Ibid., pp. 69 and 74.
234 dose from a full-mouth exam: Laws (1983), Appendix C.
234 'exclude routine full-mouth X-ray examination'. Lauriston Taylor. 'Trends, Issues and Problems in Dental Radiology', keynote address at National Council on Health Care Technology Conference on Dental Radiology, June 19, 1981, p. 8.
235 'consider going to another dentist': interview with author, August 13, 1984.

23 Out of Control

236 Atmospheric testing has added to the world's environment: Eisenbud, pp. 272, 280 and 380.
237 estimated dose to each person on earth: Eisenbud, p. 396; National Academy of Sciences, p. 66.
237 the collective dose from the fallout that will reach earth by the end of this century: UNSCEAR, 1982.
237 'the quantities of debris that remain underground after such tests are huge': Eisenbud, p. 273.
237 more than 500 underground tests: Jack Dennis, ed., p. 305; 'Study detects secret U.S. nuclear bomb tests', *San Francisco Examiner* (January 17, 1988); Environmental Policy Center, Briefing document on the Nevada Test Site, Washington, D.C. n.d., p. 3; *Science* (January 24, 1986), pp. 333–334.
237 about 20 tests a year: *Science* (January 24, 1986), p. 333.
237 closer to 400 underground tests since 1963: Dennis, ed., op. cit., p. 305 and *Science* (October 22, 1982), p. 360.
237 more than 100 nuclear weapons on or under Mururoa atoll: Dennis, ed. op. cit., p. 305.
237 approximately 500 million tons of fission products: Thomas Cochran, Natural Resources Defense Council, Washington D.C. telephone interview with author, October 30, 1987.
237 hundreds of earthquakes in the past 25 years: Environmental Policy Center, op. cit., p. 4.
237 one in ten underground tests had vented: 'Nuclear Testing at the Nevada Test Site' (1983), Cedar City, Utah; 'How Safe are desert A-tests?', *Boston Globe* (August 17, 1980), p. A2.
238 'a Challenger waiting to happen': *In These Times* (April 1–7, 1987), p. 7.
238 Chernobyl spewed at least 36 million curies of radioactivity across the world: Eisenbud, p. 381. The story of the Chernobyl disaster is taken from a variety of sources, including Eisenbud, pp. 375–388; Jones and Southwood, eds., pp. 251–274, and many articles in the magazines *Atom*, *Science*, and *New Scientist*.
238 Estimates of the number of fatal cancers: *Science* (September 12, 1986), p. 1141.
239 Premature deaths of from 35,000 to 40,000 people: 'New American research has alarming implications', *The Independent* (January 4, 1988). *The Economist* (January 30, 1988), p. 67.
239 the EPA's estimates of radiation levels were about three times too low: *San Francisco Chronicle* (May 20, 1986), p. 17.

Notes

239 detect as little as 15 per cent of the radioiodine present: *Los Angeles Times* (May 22, 1986), p. 28.

239 alarming the tens of thousands of people who had in the meantime eaten lamb: *New Scientist* (October 9, 1986), pp. 43–46; 'Contaminated lamb passed for sale', *The Observer* (March 6, 1988).

239 Zhores Medvedev, assuming the accident was well-known in the West, mentioned it: *New Scientist* (November 4, 1976), pp. 264–267; *New Scientist* (June 30, 1977), pp. 761–764.

240 'no cultivated fields or pastures, no herds, no people ... nothing': Wasserman and Solomon, p. 201.

240 'was not measured – this is not possible': Eisenbud, p. 39.

240 'the worst incident of radiation contamination': Dennis, ed., p. 378.

240 The Church Rock disaster: Accounts of the Church Rock disaster can be found in Wasserman and Solomon, pp. 177–182, and in Stanton, op. cit., pp. 14–17.

241 at least 15 accidental tailings releases: Wasserman and Solomon, p. 183.

241 about 190 million tons of tailings in the western states: Eisenbud, p. 176.

241 bottom sediments in Lake Mead were three times as contaminated: Metzger, p. 164.

241 there is no money for it: Bill Garrett, telephone interview with author November 14, 1985; Edward Swanson, telephone interview with author, March 12, 1986.

242 Altogether 115,000 gallons ... had leaked: Hilgartner et al., pp. 150–158; *Science* (August 24, 1973), pp. 728–730.

242 500,000 gallons of radioactive liquid since 1958: Wasserman and Solomon, p. 196. Robert Alvarez, letter to author, March 8, 1988.

242 Another 200 billion gallons: Ibid.

242 337 tons of uranium had somehow left the plant: *Environmental Action* (July/August, 1985), p. 29.

242 a scrap metal dealer bought part of a disused radiotherapy unit: This story is taken from various newspaper and magazine accounts, including Ian Anderson, 'Isotopes from machine imperil Brazilians, *New Scientist* (October 15, 1987); Geoffrey Lean and Louise Byrne, 'Brazil N-chiefs face disaster charges', *The Observer* (November 8, 1987); and William Long, 'Brazil Doctors Say Ignorance Led to Radiation Poisoning', *San Francisco Chronicle* (November 9, 1987).

243 a similar case, four years earlier, *New Scientist* (March 1, 1984), p. 7.

243 at least fifty in Brazil alone: Lean and Byrne, op. cit.

243 at least 27 Broken Arrows between 1950 and 1981: David Kaplan, 'Where the Bombs Are', in Kaplan, ed., *Nuclear California* (San Francisco: Greenpeace/Center for Investigative Reporting, 1982), p. 54.

244 just over one every three months: Ibid., p. 55.

244 'a reactor casualty' aboard the *Lenin*: David Kaplan, 'When Incidents Are Accidents', *Oceans* (July 1983), p. 29.

244 a still-classified paper: Ibid., p. 30.

244 Plutonium-238 has been used to supply power: Eisenbud, p. 239.

244 SNAP 9-A contained 17,000 curies of plutonium: Ibid., pp. 240 and 365.

244 its SNAP device fell into the Santa Barbara channel: Ibid., p. 240.

244 The device has never been recovered: Ibid.

244 less than one per cent of the radioactive deposits were found: Ibid., p. 241.

245 'a clearly measureable excess of uranium-235': *Science News* (October 31, 1987), pp. 278–9.

245 just over 40 space flights involving nuclear devices: *Earth Island Journal* (Fall 1986), p. 35.

245 will carry 280,000 curies of plutonium-238: Eisenbud, p. 241.

245 'this is a huge quantity': Ibid.

245 material which can 'survive pressures in excess of 2,000 psi': Karl Grossman, 'The Space Probe's Lethal Cargo', *The Nation* (January 23, 1988), p. 78.

Notes

245 'less than one chance in a million': Ibid.
245 'procedures will be developed': Ibid.

Epilogue

246 'brought about by painful experience': Quoted in Percy Brown, pp. 29–30.
246 from five to ten times too low: see discussion on pp. 246–247 of this book.
246 acknowledged that risk from radiation exposure is greater: ICRP, 'Statement from the 1987 Como meeting of the International Commission on Radiological Protection', Supplement to *Radiological Protection Bulletin*, No. 86 (1987), pp. 2–3.
246 'the requirement to keep all doses "as low as reasonably achieveable"': Ibid., p. 4.
246 'are expected to be completed by 1990': Ibid., p. 2.
247 immediately reduced from 5 rems per year to 1.5 rems per year: NRPB, *Interim Guidance on the Implications of Recent Revisions of Risk Estimates and the ICRP 1987 Como Statement* (November 1987), p. 4.
247 'a total factor of the order of two: ICRP (1987), op. cit., p. 3.
247 'a factor perhaps as high as five': *New Scientist* (September 3, 1987), p. 25.
247 from two to five times too high: Jones and Southwood, p. 158.
247 underestimating risk by a factor of ten or more: see note 2, above.
247 'worker and public representation': Friends of the Earth, United Kingdom, 'Response to ICRP's Como Statement' (London: December 1987).
248 'a problem of philosophy, morality, and the utmost wisdom': Taylor, L. S., 'Some Nonscientific Influences on Radiation Protection Standards and Practice', *Health Physics* vol. 39 (1980), p. 854.

Select Bibliography

Atomic Energy Commission and US Geological Service, *Prospecting for Uranium*, 1951.
Badash, Lawrence, 'Radioactivity Before the Curies', *American Journal of Physics*, 33, pp. 128–135, 1965.
Badash, Lawrence, *Radioactivity in America* (Baltimore: Johns Hopkins Press, 1979).
Ball, Howard, *Justice Downwind: America's Atomic Testing Program in the 1950s* (New York: Oxford University Press, 1986).
Bertell, Rosalie, *No Immediate Danger* (London: The Women's Press, 1985).
Bickel, Lennard, *The Deadly Element* (New York: Stein & Day, 1979).
Bleich, A. R., *The Story of X-rays from Roentgen to Isotopes* (New York, 1960).
Boyer, Paul, *By the Bomb's Early Light* (New York: Pantheon Books, 1985).
Bradley, David, *No Place to Hide* (Hanover, New Hampshire: University Press of New England, 1983).
Brecher, R. and E., *The Rays – a history of radiology in the U.S. and Canada* (Baltimore: Williams and Wilkins, 1969).
Brooks, B. G., 'Occupational Radiation Exposure at Commercial Nuclear Power Reactors and Other Facilities, 1984', Nuclear Regulatory Commission, Washington D.C. (November 1984).
Brown, Percy, *American Martyrs to Science Through the Roentgen Rays* (Springfield, Ill.: Charles Thomas, 1893).
Clarfield, Gerard H. and Wiecek, William M., *Nuclear America: Military and Civilian Nuclear Power in the United States, 1940–1980* (New York: Harper and Row, 1984).
Clark, Ronald, *Birth of the Bomb* (New York: Horizon Press, 1961).
Coggle, J. S., *Biological Effects of Radiation* (London: Taylor & Francis, 1983).
Committee on the Biological Effects of Ionizing Radiations (BEIR Committee), *The Effects on Populations of Exposure to Low Levels of Ionizing Radiation: 1980* (Washington D.C.: National Academy Press, 1980).
Compton, Arthur, *Atomic Quest: A Personal Narrative* (Oxford University Press, 1956).
Curie, Eve, *Madame Curie* (London: Heinemann, 1938).
Dennis, Jack, ed., *The Nuclear Almanac: Confronting the Atom in War and Peace* (Reading, Massachusetts: Addison-Wesley, 1984).
Dewing, Stephen B., *Modern Radiology in Historical Perspective* (Springfield, Ill.: Charles Thomas, 1962).
Divine, Robert, *Blowing On the Wind: The Nuclear Test Ban Debate 1954–1960* (New York: Oxford University Press, 1978).
Donizetti, Pino, *Shadow and Substance* (Oxford: Pergamon, 1967).
Ecker, Martin and Bramesco, Norton, *Radiation* (New York: Vintage Books, 1981).
Eisenbud, Merril, *Environmental Radioactivity* (New York: Academic Press, 1987).
Evans, Robley, 'Radium in Man', *Health Physics*, 27, pp. 497–510, 1974.
——, 'Acceptance of the Coolidge Award', *Medical Physics*, Vol. 11, No. 5, Sept–Oct, pp. 579–581, 1984.
Ford, Daniel, *Meltdown* (New York: Simon & Schuster, 1986).
Glasser, O., *Wilhelm Conrad Roentgen and the Early History of the Roentgen Rays* (London: John Bale and Sons, 1933).
Gofman, John, and Tamplin, Arthur, *Poisoned Power* (Emmaus, Pa.: Rodale Press, 1979).

Bibliography

Gofman, John, *Radiation and Human Health* (San Francisco: Sierra Club Books, 1981).

Green, Patrick, *The Controversy over Low Dose Exposure to Ionising Radiations* (Graduate thesis, University of Aston in Birmingham, 1984).

Groueff, S., *Manhattan Project: the Untold Story of the Making of the Bomb* (Boston: Little, Brown & Co., 1967).

Hacker, Barton, *The Dragon's Tail: Radiation Safety in the Manhattan Project, 1942–1946* (University of California Press, 1987).

——, *Elements of Controversy: A History of Radiation Safety in the Nuclear Test Program* (University of California, forthcoming).

Health Research Group and Oil, Chemical and Atomic Workers International Union, 'Petition requesting an emergency temporary mandatory standard for radon daughter exposure in underground mines', 1980.

Hendee, William R., ed., *Health Effects of Low-Level Radiation* (Norwalk, Connecticut: Appleton-Century-Crofts, 1984).

Hertsgaard, Mark, *Nuclear, Inc.* (New York: Pantheon Books, 1983).

Hilgartner, S., Bell, R., and O'Connor, R., *Nukespeak: The Selling of Nuclear Technology in America* (London: Penguin, 1982).

International Commission on Radiological Protection, *Recommendations of the International Commission On Radiological Protection*, ICRP Publication 26, Jan., 1977 (referred to as ICRP-26).

Johnson, Giff, *Collision Course at Kwajalein: Marshall Islanders in the Shadow of the Bomb* (Pacific Concerns Resource Center, Honolulu, Hawaii, 1984).

Jones, R. and Southwood, R., eds., *Radiation and Health* (Chichester, UK: John Wiley and Sons, 1987).

Josephson, Matthew, *Thomas Edison* (London: Eyre and Spottiswoode, 1961).

Jungk, Robert, *Brighter Than a Thousand Suns* (New York: Harcourt Brace Jovanovitch, 1958).

Kaplan, David, Ed., *Nuclear California* (San Francisco: Greenpeace/Centre for Investigative Reporting, 1982).

Kathren, Ronald, 'Historical Development of Radiation Protection and Measurement', *Handbook of Radiation Protection and Measurement, Vol. III*, A. Brodsky, ed. (Boca Raton: CRC Press, 1982).

——, *Radioactivity in the Environment* (London: Harwood Academic Publishers, 1984).

——, 'Early X-Ray Protection in the United States', *Health Physics*, 8, pp. 503–511, 1962.

Kevles, Daniel, *The Physicists* (New York: Vintage Books, 1979).

Lamont, Lansing, *Day of Trinity* (New York: Atheneum, 1965).

Lang, Daniel, *Early Tales of the Atomic Age* (New York: Doubleday, 1948).

——, 'A Most Valuable Accident', *The New Yorker*, pp. 49–92, May 2, 1959.

——, *From Hiroshima to the Moon* (New York: Simon & Schuster, 1959).

——, *The Man in the Thick Lead Suit* (New York: Oxford University Press, 1954).

Lapp, Ralph, *The Voyage of the Lucky Dragon* (New York: Harper & Row, 1958).

Lapp, Ralph and Schubert, Jack, *Radiation: What It Is and How It Affects You* (New York: Viking, 1957).

Laws, Patricia, *The X-Ray Information Book* (New York: Farrar, Straus & Giroux, 1983).

——, *X-Rays: More Harm Than Good?* (Emmaus, Pa.: Rodale Press, 1977).

Lean, Geoffrey, ed., *Radiation: Doses, Effects, Risks* (United Nations Environment Program, Nairobi, 1985).

Lewis, Richard, *The Nuclear Power Rebellion* (New York: Viking, 1972).

Makhijani, Arjun and Albright, David, 'Irradiation of Personnel During Operation Crossroads: An Evaluation Based on Official Documents', International Radiation Research and Training Institute, 317 Pennsylvania Ave, SE, Washington, D.C. 20003. 1983.

Mazuzan, George T. and Walker, Samuel, *Controlling the Atom: The Beginnings of Nuclear Regulation, 1946–1962* (University of California Press, 1984).

Bibliography

McKay, Alwyn, *The Making of the Atomic Age* (Oxford: Oxford University Press, 1984).

McPhee, John, *The Curve of Binding Energy* (New York: Farrar, Straus & Giroux, 1974).

Metzger, Peter, *The Atomic Establishment* (New York: Simon & Schuster, 1972).

Moss, Norman, *The Politics of Uranium* (New York: Universe Books, 1982).

Mould, Richard, *A History of X-rays and Radium* (London: IPC Business Press, 1980).

Mutscheller, Arthur, 'Physical Standards of Protection Against Roentgen Dangers', *American Journal of Roentgenology*, Vol. 13, pp. 65–70, 1925.

National Academy of Sciences, *A Review of the Use of Ionizing Radiation for the Treatment of Benign Diseases* (US Department of Health, Education, and Welfare, Washington D.C., 1977).

National Committee for Radiation Victims, ed., *Invisible Violence*, Proceedings of the National Citizens' Hearings for Radiation Victims, April 10–14, 1980.

Operation Crossroads. The Official Pictorial Record (New York: William H. Wise and Co., 1946).

Poch, David, *Radiation Alert* (New York: Doubleday, 1985).

Pochin, Edward, *Nuclear Radiation: risks and benefits* (Oxford: Clarendon Press: 1983).

Pringle, Peter and Spigelman, James, *The Nuclear Barons* (New York: Avon Books, 1981).

Public Citizen Litigation Group, *On Petition for Review to the Mine Safety and Health Administration*, March 4, 1985.

Quimby, Edith, 'The Background of Radium Therapy in the United States, 1906–1956', *American Journal of Roentgenology*, 75, 1956.

Rafferty, Keven; Loader, Jayne and Rafferty, Pierce, *The Atomic Cafe: The Book of the Film* (New York: Bantam Books, 1982).

Rippon, Simon, *Nuclear Energy* (London: Heinemann, 1984).

Rosenberg, Howard, *Atomic Soldiers: American Victims of Nuclear Experiments* (Boston: Beacon, 1980).

Saffer, Thomas and Kelly, Orville, *Countdown Zero: G. I. Victims of U.S. Atomic Testing* (New York: Penguin, 1983).

Stanton, Kathleen, *Uranium: In the atom's shadow* (Arizona Republic, 1983).

Sternglass, Ernest, *Secret Fallout* (New York: McGraw-Hill, 1981).

Stone, Robert S., 'Health Protection Activities of the Manhattan Project', *Proceedings of the American Philosophical Society*, vol. 90, no. 1, January, 1946, pp. 11–19.

——, 'General Introduction to Reports on Medicine, Health Physics, and Biology' in Stone (ed.), *Industrial Medicine on the Plutonium Project: Survey and Collected Papers* (New York: McGraw-Hill, 1951).

Taylor, Lauriston S., 'Radiation Protection Standards', *CRC Critical Reviews in Environmental Control*, April, 1971, pp. 81–124.

——, *Organization for Radiation Protection*, US Department of Energy, Washington D.C., 1979.

——, 'Judgement in Achieving Protection Against Radiation', IAEA Bulletin, Vol. 22, No. 1, pp. 15–22.

——, 'Some Nonscientific Influences on Radiation Protection Standards and Practice', *Health Physics*, vol. 39, December, 1980.

——, 'Radiation Protection Standards: What They Are and What They Are Not', Presented at Eighth Life Sciences Symposium, Los Alamos Scientific Laboratory, 9 October, 1980.

Taylor, Raymond and Taylor, Samuel, *Uranium Fever* (New York: Macmillan, 1970).

US Congress, House Committee on Interstate and Foreign Commerce. Subcommittee on Oversight and Investigations, and Senate Labor and Human Resources Committee, Health and Scientific Research Subcommittee, and the Committee on the Judiciary, *Health Effects of Low-Level Radiation*, 96th Congress, 1st session, April 19, 1979.

US Congress, House Committee on Interstate and Foreign Commerce, Subcommittee on Oversight and Investigations, *The Forgotten Guinea Pigs*, 96th Congress, 2nd session, August, 1980.

US Congress Joint Committee on Atomic Energy, *Radiation Exposure of Uranium Miners*, 1967.

US Congress, Senate Committee on Labor and Human Resources, Subcommittee on Health and Scientific Research, and Committee on the Judiciary, *Health Impact of Low-Level Radiation, 1979*, 96th Congress, 1st session, June 19, 1979.

US Senate Special Committee on Aging, *Occupational Health Hazards of Older Workers in New Mexico. Aug 30 1979*, 96th Congress, 1st session, Government Printing Office, Washington, D.C., 1980.

United States Government Accounting Office, *Operation Crossroads: Personal Radiation Exposure Estimates Should Be Improved*, November 1985 (GAO/RCED-86-15).

Wagoner, Joseph, 'Uranium Mining and Milling: The Human Costs', lecture to University of New Mexico Medical School, March 10, 1980.

——, 'An Assessment of the Adverse Health and Environmental Consequences of Mining, Milling, and Waste Disposal of Uranium-Bearing Ores', Statement to New Jersey State Senate, Environment and Energy Committee, Jan. 20, 1981.

Weart, S. R. and Szilard, G. W., eds., *Leo Szilard: His Version of the Facts: Selected Recollections and Correspondence* (Cambridge: MIT Press, 1972).

Index